FROM APES
TO ASTRONAUTS

ADRIAN BERRY

The Daily Telegraph

ACKNOWLEDGMENTS

Grateful thanks are due to John Delin and Keith Heron for their help in preparing this anthology, both of whom made many useful suggestions for improvement.

CONDITIONS OF SALE:
This book is sold subject to the condition that
it shall not, by way of trade or otherwise, be lent,
re-sold, hired out or otherwise circulated without
the publisher's prior consent in any form of
binding or cover other than that in which it is
published

First published 1980

© *Daily Telegraph* 1980
ISBN 0 901684 60 0
Printed in Great Britain
by Purnell & Sons Limited,
Paulton (Bristol) and London

Contents

The Science Writer
By Isaac Asimov

You might suppose that a science writer—especially one who is a correspondent for a daily newspaper—is much in the same position as a sports writer, or the fellow who runs the society column (if there is such a thing anymore) or the drama reviewer. The science writer, it might seem, is just another one of those who interprets a specialised field for the interested but inexpert onlooker.

Yet there must be more to it than that, simply because there's more than that to science itself.

It's quite possible that there may be far more people who wonder whether the new play that's opened is any good or not, than ever wonder about any phase of science; and far, far more than either who would like to know who's going to win tomorrow's soccer game—but importance is not the sort of thing that can safely be left to majority vote.

A drama misjudged, or a soccer victory unforeseen, may be the cause of much anguish, and even some monetary loss, but if science goes wrong these days, the result might just possibly be an appalling catastrophe before which the most expensive of wasted tickets and lost wagers must pale.

Again, the world faces an appalling list of life-and-death crises these days; life-and-death for our entire industrial civilisation and, not so incidentally, for some billions of those who make it up and who (both themselves and in the person of their recent ancestors) have flourished and multiplied under it for over two centuries.

The sad litany is so well known that it is scarcely worth going through it—dwindling resources, including, most particularly, oil; gathering pollution; increasing weight of over-population; vanishing topsoil and expanding deserts; multiplying nuclear weaponry; deepening pin-pricks of terrorism and social alienation; intensifying popularity of various cults of the irrational.

What solutions to these can we imagine? Perhaps none, but if there are to be any at all, they will arrive through advances in science and technology which can (or, at least, *might*) supply us with new energy sources, better means for recycling and conservation, truer understanding of reproductive physiology and psychology, more efficient means for detoxifying the environment; and so on.

If science and technology are allowed to advance, that is.

It is by no means certain that they will be. A great many of Earth's problems arise from the previous success of science and technology because that very success made unwise excesses so easy and attractive. Technology has too often been driven hard by the short term goals so clearly in view, to the (much less visible) long-term detriment of humanity. And the very people who cheered on the folly as long as they profited, now turn savagely on science when the consequences come to be weighed and measured. What else is there to blame? the mass-greed and mass-ignorance of millions?

It is easy, in any case, to feel disillusioned about science—which is rational, cold, and cautious by its very nature, and is only able to tell us that it seems likely, on the evidence, that two and two is four. How much more fascinating and comforting are the fringe-beliefs which assure us, in the most confident and heartening manner possible, that two and two is certainly equal to six and a half.

Under these circumstances, it might appear that humanity doesn't have a hope.

If, as it seems to me, the only reasonable chance we

have of pulling out of the appalling hole into which we have dug ourselves is to encourage scientific and technological advance and use that advance with measured caution and judicious wisdom . . .

And if the record of the last couple of centuries is such that we seem much more likely to use the advance, as always, for immediate short-term comfort and profit, without consideration for the long-term consequences . . .

And if there is, in any case, growing disillusionment with this miserable record and, in our compounded folly, we turn away not from our own unwisdom, but from the principle of science and rationality . . .

Then, surely, we face certain ruin.

Except that it goes against the grain of human beings to give up and accept defeat. Perhaps, if those of us who can, patiently explain the principles of science, describe the latest advances, point out the fallacies in the silly pap that is fed the public by cleverly irrational knaves (or sincere ignoramuses), then enough people might be swung to the side of rationality and wisdom to save us yet.

In 1920, H. G. Wells said, "Human history becomes more and more a race between education and catastrophe" and in the sixty years that have passed since, education seems to be losing the race. Ironically, as our growing scientific and technological capacities make that, more and more, the likely agent of catastrophe, it is scientific education (of all things) that falls farthest behind.

The more reason to redouble our efforts!

The science writer, especially the science-correspondent of a great newspaper, faces the most difficult and the most important part of that task. He faces the public, the great mass of humanity, far more than any academic in the classroom does or can.

The science writer has an audience that can, all too easily, skip his essay and turn to the advice-to-the-lovelorn column. And if he can manage to seize some of that audience, he must often tell them what they don't want to hear—that what they would like to believe in is nonsense that is leading them astray.

Yet he *must* do all this, and if he can succeed even a little bit, we are that much farther from the edge of the cliff.

It is consequently heartening, at this point, to be able to present a collection of Adrian Berry's essays from *The Daily Telegraph*. Adrian knows what he is saying, he writes well and interestingly, he makes matters plain, and he is fearless in assaulting the forces of irrationality and superstition.

And he is on the side of the angels—for reason, for the advancing front of knowledge, for survival and growth, for the magnificent heights of the future.

Isaac Asimov
New York
July, 1980.

To Linda

EVOLUTION

Despite overwhelming evidence that the Earth is nearly 5,000 million years old and that life has existed on it for three quarters of that time, an astonishingly large number of people still believe literally in the story of Genesis as told in the Bible. Yet the Scriptures are quite incompatible with science; they depict the Earth as having been created only 4,000 years ago, less than a millionth of its true age. The first article in this book, published in June, 1978, provoked nearly 300 letters of protest from indignant fundamentalists.

When Man makes a monkey of his past

MANKIND'S ability to survive has always been a pressing topic. Since this month marks the 144th anniversary of Charles Darwin's entry into the Pacific in the Beagle, and the 118th anniversary of the row in Oxford over man's ancestry between Thomas Huxley and Bishop Samuel Wilberforce, and approaches the 53rd anniversary of the Tennessee monkey trial, it might be an appropriate time to comment on the current state of the theory of evolution and view our past survival.

For the crime of suggesting that man and ape have a common ancestor, Darwin was probably more abused than any scientist who has ever lived. "Rotten fabric of speculation. . . . Utterly false. . . . Deep in the mire of folly. . . . I laughed till my sides were sore." These were some of the remarks that greeted publication of his "On the Origin of Species" in 1859.

The furious debates that followed this best-seller (the first edition of 1,250 sold out on the first day) produced at least two instances of memorable repartee. The first occurred at a packed public meeting at Oxford, when Bishop Wilberforce, known as "Soapy Sam" from his habit of constantly rubbing his hands, made a prosecuting speech with the intention to "smash Darwin."

He was confronting Thomas Huxley, a crusading evolutionist and a far less polite man than his friend Darwin. Part of their dialogue may be familiar to many people but is worth repeating:

WILBERFORCE: *"Do you truly believe that you are descended from an ape? It would be interesting to know in that case whether the ape in question was on your grandfather's or your grandmother's side."*

HUXLEY: *"I would rather have a miserable ape for a grandfather than a man of great influence who employs that influence for the mere purpose of introducing ridicule into a grave scientific discussion."*

Sixty-five years later, in the small town of Dayton, Tennessee, another remarkable clash took place. A schoolteacher had been prosecuted for defying the State's new law that forbade the teaching of Darwin's theory. The populist politician William Jennings Bryan undertook the prosecution, and Clarence Darrow appeared for the defence.

The schoolteacher was convicted of breaking the law and fined a nominal sum, but the trial, with its cross-talk of Darrow and Bryan about literal understanding of stories in the Bible (splendidly reproduced later in the film "Inherit the Wind," starring Spencer Tracy and Frederic March), caught the attention of the world.

Darrow drew attention to radioactivity in rocks, which showed that the earth was thousands of millions of years old, which it needed to be for Darwin's theory to be true.

DARROW (producing a piece of rock): *"Have you any idea, sir, of the age of this rock?"*

BRYAN: *"Sir, I am more interested in the Rock of Ages than the age of rocks."*

The dispute continues to this day. While the principle of evolution is now accepted unanimously by serious scientists, education authorities throughout the Western world are at any time liable to receive demands from militant fundamentalists that they allot "equal time" to the teaching of evolution and to the Biblical version of the creation.

A few, to their shame, submit, allowing themselves to be intimidated into teaching their pupils a version of the origin of mankind, and indeed of all life, whose chances of being true are virtually zero.

Evolution is a simple idea, and evidence for the truth of it can be found by a visit to any zoo. The keys to it are "natural selection" and the "survival of the fittest." When the environment changes for the worse, all animals die except those who can adapt themselves for survival.

To take an obvious example, giraffes have long and blotchy necks. The reasons for this are plain. Those with short necks could eat fewer tree-leaves than their brethren and died out, while those with uncamouflaged necks were easily spotted by predators.

Nature thus selected the fittest giraffes for survival; and over time-scales of hundreds of millions of years, every other life-form on earth, including man, survived in the same way through being the fittest and most able to adapt.

I should add that the once derided alternative possibility that species *adapt* to meet changing environments, rather than being wiped out if they failed to adapt, is again being seriously discussed in the light of curious colour changes of certain moths and in the coats of some wolves and Arctic foxes.

Yet there is no alternative to the main theory of evolution. Anyone who rejects it must believe either in a supernatural event, or else that the planet was long ago colonised by humanoids from another planet who, for some strange reason, brought with them the ancestors of *all* the animals. There is no scientific evidence for either of these propositions.

Why, then, does evolution still have so many vociferous opponents? The answer, may be, as Prof. Robert Jastrow remarks in his excellent new book "Until the Sun Dies" (Souvenir Press, £3·95), that these are essentially unimaginative people who lack the intellectual ability to think about time-scales of hundreds of millions of years and, like Bishop Wilberforce, they feel insulted at the thought of being cousins of the apes.

A God who reveals himself

THE comments of nearly 300 readers on my column, "When Man makes a monkey of his past," pointing out that the theory of evolution gives a more convincing account of man's origins than the Book of Genesis, are an encouragement to search for some common ground between the rival claims of religion and science.

"Arrogant . . . dogmatic . . . the work of a cowboy . . . discredited Darwinian rubbish." These were some of the more extreme comments. Others were more thoughtful, especially about that ultimate mystery, the nature and identity of God.

"But who and where is God?" asked one reader, Mr F. Bean. The Rev. John R. Berry replied: "This question is rather akin to saying 'Where is gravity?' or 'Where is electricity?'" "God has simply used science, through cosmology and evolution, to create man," answered Mr S. Harris.

Cosmology and evolution. The laws of gravitation and mass-energy on one side, and those of biochemistry on the other. It ought to be possible, as Mr Harris suggests, to construct an image of the Creator which does not conflict with scientific observation.

It is difficult to believe, with Genesis, that God directly and in person created the Earth in the space of six days.

An actual solar system is in the process of formation 10,000 light-years away in the constellation of Cygnus the Swan. The creation of this solar system so far out in space may be the mirror image of what happened to the Sun and its planets some 5,000 million years ago. It is a very different picture to the one given by Genesis.

One can believe in God, or a supreme being, without

having to take too seriously the more fabulous stories of the Old Testament, a document whose connection with Christianity is extremely tenuous.

By what technological devices, for instance, did Joshua cause the Sun to stand still in the heavens, or Methuselah live 969 years, or Elijah fly up in a fiery whirlwind to heaven, or Jonah live for three days, singing psalms, in the belly of a whale, or Eve take shape from one of Adam's ribs?

It was not, of course, physically impossible for Joshua to have halted the Earth's rotation. But from Isaac Newton's First Law, that "a body moves in uniform motion unless acted on by a force," we may estimate that the energy needed to do it would have amounted to the equivalent of many millions of large hydrogen bombs. Similar objections apply to all other biblical "miracles."

Many scientists, rejecting these tall tales but still wanting to believe in a supreme intelligence, take their cue from Albert Einstein's reply when challenged about his religious beliefs:

"I believe in a God who reveals himself in the harmony of all that exists, but not in a God who concerns himself with the fate and actions of men."

A lecture on this theme was given by the astronomer Prof. Robert Jastrow at this year's meeting in Washington of the American Association for the Advancement of Science, who announced he was setting out on a personal search for evidence of God, but admitted that the odds against his success were "virtually insurmountable."

Yet Prof. Jastrow and his fellow-astronomers have one great advantage over the philosophers and theologians. They know where to start looking: namely at the Big Bang, the primordial fireball from which the universe was created, together with space and time, some 18,000 million years ago.

This idea seems intelligent and practical, for it was the Big Bang that created the unbreakable laws of nature, Mr Harris's "cosmology and evolution" which have since governed the universe.

Every event has a cause, Prof. Jastrow argued. It should therefore be possible to look backwards in time beyond the Big Bang, and seek a clue to whether this cataclysmic event had a cause in the form of some intelligent design.

How can we look backwards in time for 18,000 million years? Very simply, by using the telescopes. Since light and radio waves have always travelled at a constant speed, the further we look into space the further back we are looking into time.

At the very edge of the universe are the still detectable traces of the Big Bang, a continuous radio static coming from all parts of space that was discovered in 1965 by two telephone engineers.

Yet a strange difficulty presents itself, as if God for some reason did not wish us to examine too closely the mysteries of the Creation. The Big Bang itself is hidden by "cosmic censorship." Because the primordial fireball was infinitely hot and infinitely dense, the electrons and the nuclei of the first atoms were separated, and matter was thus rendered opaque.

The only solution might be for deep-space astronomers to study those sub-atomic particles called neutrinos, the messengers which penetrate every obstacle, and might still carry to us further information about what happened in the very beginning.

If the God postulated by science seems a less satisfactory being than conventional deities, we may take some comfort from the hymn couplet:

Laws which never shall be broken
For our guidance He hath made

At least the astonomers' approach seems preferable to Nikita Krushchev's frivolous explanation of the Soviet space programme: "We sent up Yuri Gagarin to see if he could find the Kingdom of Heaven, and he couldn't see it. So we sent up Gherman Titov to make sure. And he couldn't find it either."

The creationist "case"

BIBLICAL fundamentalism, the belief that Old Testament stories of the world's beginning are true in every detail, is a phenomenon I had expected to find today only among extremists or hillbillies. But in mixed disbelief and alarm I recently attended a conference at Central Hall, Westminster, of at least 550 of these true believers.

All of them rejected evolution, and most believed emphatically that the world, complete with Adam and Eve and all the animals, was created in precisely six days. I will now present the "case" of these "creationists," or fundamental believers, adding my own comments when necessary.

1. *The earth is only a few thousands of years old—it must be if the story of Genesis is literally true. The record of radioactivity in the earth's rocks, which show it to be 4,600 million years old, are admittedly unreliable and therefore false.*

2. *So young is the earth that dinosaurs were contemporary with men.* Speakers at the conference claimed that a cave drawing existed of a large dinosaur, drawn by an early man, who was obviously depicting what he had seen. I have seen photographs of many cave drawings, but I have never seen this one. I would lay a bet that I never will.

3. *In the pre-Cambrian rocks, more than 500 million years old, no fossils have ever been found.* But this does not satisfactorily explain why I keep seeing pictures of them in scientific journals. The *Scientific American*, in a recent issue devoted entirely to evolution, reports of some jellyfish fossils in Australia about 680 million years old—but it's no use my talking about them since the creationists say they don't exist.

4. The evolutionary link between small dinosaurs and birds is archaeopteryx, the tiny winged dinosaur that lived about 100 million years ago. He had half-developed feathered wings, useless for flying. *So what were his feathers for if he couldn't fly?* Scientists say for insulation, but this explanation, according to the creationists, is "incredible."

5. *The earth's magnetic field is weakening steadily. Ten thousand years ago it was so strong that the earth (if it existed) would have been uninhabitable.* I had never heard this one before.

6. *All fossil records of early man are fakes. Piltdown Man was a fake, therefore all the others were fakes. That proves it, doesn't it?*

7. *The Flood occurred exactly as stated in Genesis. It covered the entire earth. Supporting stories about a flood appear in the ancient writings of many peoples.* Actually, these people must have been writing about different events, since the floods reported are often separated from each other by centuries.

Perhaps these ancient writers had their dates muddled. Yet despite these errors, we are told to accept their word for a world-wide flood despite the lack of any geological evidence.

8. *Noah's Ark was constructed precisely as stated in Genesis 6.* This remarkable vessel was 500 ft long, 83 ft broad, and 50 ft high, taking a cubit as 20 inches. Without a keel, so top heavy a ship would have capsized at the first gale. But perhaps supernatural ships don't need keels.

9. Mr Brian Grantham-Hill, head of biology at Torquay Girls Grammar School, told the conference: "The Bible is accurate from cover to cover. When it says the Creation took place in six days, it means six days." It also says that Eve was made from Adam's rib, but who am I to argue with a head of biology?

10. *Lord Kelvin's Second Law of Thermodynamics predicts that disorder in the universe is continuously increasing as heat (thermal energy) decreases. This law*

must surely rule out evolution which seems to predict the opposite, an increase in order. Well, not exactly for the Second Law deals only with "isolated systems," like single creatures or single stars.

Each of us must grow old and die, but there is nothing to prevent a species from surviving and evolving into a sub-species or to a new species altogether.

11. *Evolution is only a theory after all, a mere hypothesis, and is no better than any other.* But in science these two words mean different things. A hypothesis is a provisional conjecture about causes and relationships, while a theory is a *verified* hypothesis.

The 1977 McGraw-Hill Encyclopaedia of Science and Technology is dissatisfied even with this dictionary definition. "Formerly regarded as expressing simply a theory," it says, "organic evolution is now an integral part of any modern biological synthesis, as a 'fact' in the main body of science."

Or, as Professors Sir Fred Hoyle and Chandra Wickramasinghe state in their new book "Lifecloud": "Darwin's theory is now accepted without dissent."

Despite the large audience at last weekend's conference, one must remember that the creationists are a tiny minority among Christians who, generally speaking, have no objection to Darwin's theory.

How evolution works

RELICS have been found of an unpleasant-looking ape-like creature who is believed to have been our direct ancestor. He lived about 30 million years ago in a region of lush forest country that is now the Sahara Desert and which is not far from Cairo.

He is the oldest known common ancestor of both

apes and people, and some of his bones were recently extracted from the sand by a team of anthropologists from Duke University, North Carolina.

How can we be sure that he was the ancestor of man and not merely of some species of ape? The answer is that we share with him a sufficient number of *homologies,* or common characteristics, for our descent from him to become highly probable.

The eye sockets of this "dawn ape" indicate that he was a diurnal creature, active mostly in daylight. This suggests that he lived in groups, and probably employed some primitive form of speech. This implies in turn a degree of social behaviour which would have forced him to be more competitive and courageous than solitary animals.

Other evidence bears out these speculations. The males had large, fang-like canine teeth, no doubt for defending their territory and their females. And he seems to have been the most intelligent land animal in the world at the time. His cranial capacity was about $1 \cdot 8$ cubic inches, which was larger, relative to body size, than any other contemporary mammal.

One's first reaction to this dawn ape is one of disbelief that this 12-lb creature could have fathered a race which split the atom and walked on the moon. One cannot suppress a feeling of wonder that such tremendous steps forward could have occurred within 30 million years—a relative flash of a second in astronomical or geological time.

The case of the dawn ape demonstrates an important technique in evolution science: how to show, by comparing characteristics in common, that different species have an identifiable common ancestor.

Species that appear at first to be wholly different can turn out to be more closely related to each other than to more similar-seeming animals.

Consider the salmon and the squirrel. Which is more closely related to the dolphin? Most people would instinctively say the salmon, since both salmon and dolphins, unlike squirrels, are sea-going and stream-lined. But in fact the squirrel is the dolphin's closer relative.

Why? The dolphin's characteristics tell the story. It has two sets of teeth, a four-chambered heart, a streamlined shape, mammary glands, and it breathes air. The squirrel shares four out of five of these homologies, and the salmon only one. Therefore, the dolphin and the squirrel had a common ancestor which lived much more recently than the common ancestor of the dolphin and the salmon.

Animals that merely look alike are often mistaken for close cousins. The tortoise and the armadillo, for example, both walk slowly under protective shells. But a comparison of their shells shows their differences. The ribs of the tortoise form part of its shell, but the armadillo's ribs do not. Hence, they are far apart in evolution, and only look alike because their shells have a common function.

Biblical fundamentalists and others who lack the imagination to comprehend the theory of evolution often reject it because they fail to see any evolutionary changes actually happening. They do not understand that such changes take place very slowly and over very long time-scales. Consider the following six words:

DICE
DINE
DIRE
DARE
CARE
CARD

Each differs from the last by only one letter, yet the changes are cumulative. The words DICE and DINE are very similar, while DICE and CARD differ from each other as much as we differ from the dawn ape.

This analogy is one of several I have drawn from an excellent new picture-book on evolution, "Dinosaurs and their Living Relatives" (hardback edition £9, paperback £2·95, published jointly by Cambridge University Press and the Natural History Museum, South Kensington). This book should make relationships clear even to the most stubborn anti-evolutionists.

The domineering gene

Descended from monkeys? My dear, let us hope that it isn't true. But if it is true, let us hope that it doesn't become widely known!
> —The wife of Bishop Wilberforce in 1860.

ANY discussion of evolution seems to provoke letters of complaint from people who are infuriated at the thought of being descended from animals. Perhaps still more annoying will be the inference that we ourselves are "animals," and that we behave as animals throughout our lives.

The theory of evolution becomes more interesting still when it concerns not merely our physical descent—the way in which our bodies have evolved from more primitive creatures—but the way we think and behave. Such a science is the relatively new sociobiology, or "evolutionary biology."

Philosophers have long sought a means to explain and predict human behaviour, and lacking the information provided by Darwin, they got most of their ideas wrong. They took extreme positions. Either all instincts and emotions were acquired by life experience, or else everything was inherited.

David Locke, in the 17th century, took one such view. The infant human mind, he declared, is a piece of "white paper, void of all character, without any ideas. How comes it to be furnished? To this I answer, *Experience.*"

Two centuries earlier, King James IV of Scotland had tried with a cruel experiment to prove the opposite. He isolated several children at birth, and forbade anyone to speak to them. He was convinced they would develop an innate God-given language, which he piously hoped would be Hebrew. Not surprisingly, they all died without learning to speak.

Sociobiology demolishes these bizarre ideas. Its main prediction is that our emotions are governed overwhelmingly by the desire that those who share our genes, namely our blood-relatives, shall survive in sufficient number; and that this desire is inspired less by altruism than by the "selfishness" of the genes themselves.

Like many new theories, it sounds both extraordinary and unprovable. But as a remarkable new book shows, it has been tested on animals and on primitive tribes. In "Sociobiology; the Whisperings Within," (Souvenir Press, £6·50) David Barrash, a professor of psychology and zoology at the University of Washington at Seattle, overcomes our disbelief.

The great biologist J. B. S. Haldane, Prof. Barrash tells us, was once asked whether he would sacrifice his life for his brother. Haldane did a swift calculation. No, he replied, not for one brother. But he would sacrifice himself either for *three* brothers, or for *nine* cousins.

His reasoning, he explained wryly, was based purely on biology; humans share one half of their genes with siblings, one quarter with half siblings, and one eighth with cousins. So any gene that influenced its carrier (himself) to risk its body to save three brothers (or nine cousins) would result in making more copies of itself than would be lost, even if the individual died in the attempt.

We might raise the objection that genes are embodied in the human cell, which is hardly large enough to contain the pocket calculator which these computations might require.

There is no such need, says Prof. Barrash. "Evolution has done the arithmetic during the many long generations of the history of every species. In the course of time, some genes have directed their bodies to make bad choices. They made errors in solving the critical equation that includes costs and benefits. These error-prone genes have left fewer descendants than those whose calculations were more accurate.

"We might do poor arithmetic in school, but we and everything else that lives behave like mathematical geniuses. We are selected to do the right things although we do not know why."

This strange notion nevertheless strikes a chord. Most of us know that we would unhesitatingly sacrifice our lives to save our children.

By the same token, we view with particular horror the "unnatural murder" by parents of their children or of other close relatives. The macabre thrills in Shakespeare's plays "King John" and "Richard III" are provided by the royal murder of nephews. One suspects that these plays would not be nearly such good box office if the two kings had merely murdered somebody else's nephews.

In contrast, the wicked step-parents who persecute their spouses' children are essentially comic stock-characters in the story books. Their crimes excite no particular revulsion. After all, they have no genetic investment in their victims.

An early proponent of sociobiology, Prof. E. O. Wilson, draws attention to the phenomenon of instincts inherited from thousands of generations. "People everywhere," he points out, "at a very early age have already developed a deep horror of snakes and spiders with nothing more than a gentle nudging from their parents.

"Yet although parents constantly discourage their children from going near electric sockets, cars, knives and the like, phobias against such objects rarely develop."

Sociobiologists hope their theory of behaviour, which seems at present to apply only to individuals, can be expanded into a science which could analyse humanity as a whole.

SPACE TRAVEL

The great Russian space pioneer Konstantin Tsiol-
kovsky once remarked: "Earth is the cradle of mankind, but one cannot stay in the cradle for ever."
Regular manned flights of the American space shuttle
are likely to mark the true beginning of the colonisation of space.

Shuttle service into space

OF ALL the technological enterprises on which man has embarked in the past decade, the one which seems to me the most essential and which offers the highest hopes for the future of the race, is the United States space shuttle.

Now a spaceship which can only hold five people (or 10 in an emergency) and can only remain aloft for a maximum of 30 days may seem a strange candidate for offering "the highest hopes for the future of the race."

But the shuttle, of which five are initially being built, is only the beginning of an entirely new and very much cheaper form of space travel than the mighty space rockets which have hitherto thundered off the pads of Cape Canaveral and Baikonur.

The Saturn Five rocket which carried men to the moon, with nearly eight million pounds of thrust in its first stage, and its combined horsepower equivalent of 534 jet fighters, was rightly considered a creative triumph.

To watch a Saturn Five lift-off at a safe distance of three miles, to feel the spectator's pavilion tremble at that deafening roar, to *see*, but not to hear, people all around yelling "Go! Go! Go!" was an experience not to be forgotten.

But there was one snag with the Saturns and all the other rockets; when the first and second stages were exhausted they were dropped uselessly into the sea. Rockets are uneconomic.

Imagine making a long journey by air and then being told that each fare would be millions of dollars because the aircraft had now been used and would have to be thrown away. Each manned rocket flight cost hundreds of millions of dollars because the rocket could never be used again.

All the hopes of the shuttle rest on the fact that it is reusable; nothing need be thrown away. Each shuttle can fly up into space and down again hundreds of times with little expense beyond the fuel of its engines and the salaries of its crew.

Even the booster engines which, while not part of the shuttle, help it to reach orbit will not be destroyed. They will be parachuted back to earth to be used over and over again. The space shuttle will thus reduce the costs of launching any object into space by about 70 per cent, and it is for this one reason that this stubby little space vehicle is mankind's hope of a gateway to the stars.

The immediate practical importance of the shuttle lies in solving problems of the Earth. Sources of pollution will be pinpointed, new supplies of water to irrigate deserts will be found, geological discoveries will make the prediction of earthquakes much more accurate.

The oceans will be searched for new fish supplies. Maps will be revised. The secrets of our weather and climate may be unravelled. All these feats can be carried out far more efficiently from space than by aircraft, because so much larger areas can be seen.

Large amounts of this work have been done already; tens of thousands of photographs of the Earth have been taken through many different camera filters from satellites and manned stations like Skylab.

But so much yet remains to be discovered. The shuttle, with its cargo hold to carry manned stations like Europe's Spacelab, will from 1981 onwards be able to so increase our knowledge of the world that the things we have learned already since the space age began in 1957 will seem the merest introduction.

Yet all these applications will at length seem mundane and parochial. Bigger shuttles will eventually be built, with room for perhaps 100 passengers or more. Not just space stations, but permanent space colonies will be established.

Solar reflecting mirrors will one day be constructed, to beam down the sun's radiation to earth, solving for ever the energy crisis which is threatening to cause so much havoc by the end of the century.

Interplanetary and interstellar spaceships, vehicles far too massive and powerful ever to be launched safely from the earth's surface, will be constructed in space from lunar materials and asteroid fragments.

Today's embryo space shuttle, a tiny model of the much larger craft which will surely follow, represents the only path by which our descendants can escape ruinous confinement to one world. It thus offers the highest hopes for the future of our race.

President Carter and space

YEARS of indifference and neglect since President Carter arrived in the White House appear at present to have given the Russians a strong lead in space technology.

Indeed, the only achievements in space which most people associate with Jimmy Carter are to have abandoned the Skylab space station and to have mistaken the planet Venus for a flying saucer.

These jibes are only slightly unfair. Sunspot activity has been swelling the earth's atmosphere and increasing air drag on the 80-ton orbiting craft, so that it becomes increasingly difficult, but not yet impossible, to keep it aloft.

But Mr Carter peremptorily ordered space officials to cease their efforts to save Skylab, on the grounds that it would cost a few million dollars, ignoring the thousands of millions it has cost to construct, launch and maintain it.

Venus is so bright that it takes only a little upper

atmosphere turbulence to give it a monstrous and frightening appearance. An ordinary citizen unschooled in elementary science might indeed look up at Venus and imagine he has seen a flying saucer. Such a person, if sufficiently naive, might go further and announce the fact to the media, but is he fit to be in charge of an advanced industrial nation?

Mr Carter's second space budget was as sparse and unimaginative as his first. At £2,362 million, it represented a mere 7·2 per cent increase on last year's and did not even pretend to keep pace with America's present 9 per cent inflation rate.

The space shuttle, whose construction was ordered by President Nixon, is now funded and should make its maiden voyage in 1981. But the development of many of its planned ancillary systems such as the power module for extra electricity have been deferred by Mr Carter.

According to Mr Robert Hotz, editor of the authoritative journal *Aviation Week and Space Technology*, "there is no fiscal groundwork in this budget to develop the full capabilities of the shuttle as an operational system.

"Without the power module, its useful capabilities will be severely limited, and many of the bright promises now being made to industry for shuttle commitments will remain unfulfilled.

"But it is not just the prosaic budgeting that is causing concern. It is rather the lack of appreciation of the opportunities in space, and of an understanding of how they will stimulate many facets of the national economy outside the aerospace sphere."

The Russians, meanwhile, are pressing ahead in space with a constancy that the Americans have not shown since the build-up for the Apollo moon missions. Despite many disasters, they have reached the level of a semi-permanent manned space station that can be re-supplied with cargo through twin airlocks. They also are building a re-usable space shuttle.

It is true that the inflationary crisis has required many economies in US Government spending, just as it has in Britain. But the arrogant mistake of Mr Carter has been to assume that space technology is just another "down-the-drain Federal expenditure." It is rather as if a domestic householder in hard times were to try to make ends meet by abandoning his children's education.

For space is a frontier as important as any on earth. The material resources of the moon and asteroids, the vast opportunities for industrial investment and colonisation, as well as the grimmer possibilities for sophisticated war machines, make it essential that the conquest of "this new ocean," as President Kennedy once called it, should not be left to the Communists.

Fortunately, there is every hope that Congress will challenge this pusillanimous policy. Senators and Representatives are pointing out that the technology already exists to build large-scale structures in space cheaply, and that nothing is preventing them except the earthbound prejudices of Mr Carter and his Georgia backwoodsmen.

The President's attitude is understandable, if deplorable. Although by 1985 the number of people employed in space-related industries may be triple that at present, little of this great boom has been apparent during the 1980 Presidential election. We must hope that the next American President will be someone whose general scientific understanding noticeably exceeds that of a suburban sewing circle.

Moon men

A decade has passed since Neil Armstrong climbed down the ladder of the lunar craft Eagle and uttered what will surely be seen by posterity as an accurate prophecy: "That's one small step for a man, one giant leap for mankind."

An accurate prophecy? Many people today would doubt it. The pace of manned exploration of space appears to have slowed significantly since Eugene Cernan and Harrison Schmitt left the moon in 1972.

The slow-down, caused in part by the loss of public interest, was of course inevitable after such a tremendous climax as the first moon-landing and could have been predicted by any competent crowd psychologist.

This was not apparent to most space writers at the time, including myself, who believed there would be great progress in the 'seventies. By 1975, we thought, there would be large American manned space stations in earth orbit, the nuclei of future orbital cities, and that by 1980 the first permanent moon base would be in an advanced state of forward planning.

But these enthusiastic predictions, one must now admit, were based on a too hasty reading of history. Technology does not advance in a continuous progression, but in a series of starts.

History can give the impression that once a breakthrough has been made, exploitation follows rapidly. But this is an illusion.

It comes as a slight shock to realise that 128 years separated the first voyage of Columbus and the sailing of the Mayflower, that nearly a half-century elapsed between the invention of the telephone and its general availability, that the first commercial atomic power station came on line 39 years after Lord Rutherford split the atom, and that there were 19 years between the first non-stop transatlantic flight and the flight of Orville Wright.

From these precedents, we see that there is nothing very surprising at the delay between Neil Armstrong's walk on the Bay of Tranquility and the construction of the first permanent colony on the moon.

Yet there is no physical reason why hundreds of

thousands of people should not make their permanent homes on the moon, and strong reasons why they are likely to do so.

Let us look forward 50 years. Long before that time a large amount of industrial activity will be taking place in space. The shuttle and its successors will carry people and constructional materials into space for a fraction of the real cost of the voyages of the first astronauts.

But there will be limits to the weight and bulk which even the largest shuttle can carry into orbit—when we remember that a speed of 25,000 mph (7 miles per second), the escape velocity of the earth, is necessary to achieve orbit.

But this minimum speed can be reduced by a fifth with a consequent energy reduction of 97 per cent! Instead of taking large quantities of material from the earth's surface, spoiling the environment by mining and risking catastrophic accidents, why not take it instead from the moon, whose escape velocity is only 5,000 mph?

One of the most brilliant solutions to the long-term energy crisis is the idea of the American engineer Dr Peter Glaser to erect in space solar energy reflectors several miles wide, and place them 22,300 miles above the equator so that they would always remain in the same position.

They would then beam down energy in the form of radar waves directly into our national grids. A hundred or so of these reflectors could electrify a large part of the earth. Their existence could remove the main difficulty which has held back solar energy development—the problem of night and clouds; for they would be in a place where there is sunshine 24 hours a day and 365 days a year.

Constructional materials to create thousands of square miles of solar reflecting material will be available on the lunar surface. The surface rocks of the earth and moon are remarkably similar. The Moon rocks returned by the Apollo astronauts contained traces of nearly every chemical element except hydrogen.

And there is plenty of oxygen on the Moon. It has been estimated that 75 per cent of the Earth's and Moon's crust, down to a depth of 10 miles, consists of compounds of oxygen and silicon.

Astronauts of the Apollo 14 mission in 1971 found traces of rust, indicating the presence of water deep beneath the lunar surface. Even if this theory is unconfirmed, it would not be prohibitively expensive to import from Earth hydrogen, the lightest element, where it could be recycled over long periods for drinking water and for industrial purposes.

The low gravity of the Moon and the perfect vacuum of its surface offer almost limitless industrial opportunities. In a world with a surface area the size of Africa, excavated underground chambers will offer excellent living conditions for countless people, large areas under protective domes will be suitable for agriculture, and in the absence of an atmosphere all electricity can come from solar energy. Then indeed Neil Armstrong will have been a prophet.

Providing your own gravity

THE spectacle of two exhausted Russian astronauts, after their recent return from six months in orbit, having to be carried from their capsule into reclining chairs, complaining of the hardness of feather beds, and barely able to speak coherently, gave an impression of the effects of spaceflight on the human body which is wholly misleading.

All these debilitating effects, which were suffered in varying degrees by Russian and American astronauts on their return from weightlessness to the one g

surface gravity of the earth, were caused by the smallness of their spacecraft, and in the fairly near future will no longer need to be endured.

In a sufficiently large spacecraft, not only will astronauts be able to enjoy gravity of any strength they desire; they will also be able to vary the gravity in different parts of the ship, perhaps with weightlessness in their workshops and one *g* in their living quarters.

Even some of the best science fiction writers do not understand how easily gravity can be created or changed. In Alan Dean Foster's paperback version of the film "Alien," the spaceship captain gives the curious order: "Switch on the artificial gravity." Gravity then appears, as if generated by an electric current.

This is fantasy. "Artificial gravity" would have no more meaning than "artificial oxygen." Either there is gravity, or there is no gravity.

Let us make up a story of two imaginary astronauts called Charles and Carla, who are sent up for six months in a small orbital craft to take astronomical photographs and to test the effects of weightlessness on their bodies. But unlike the obedient Russians, they have no intention of obeying orders. They are happy to photograph the stars, but they decide to smuggle up in their clothes the means to introduce one *g* of gravity into their spacecraft.

They secretly purchase their gravity-producing equipment for less than £100. It consists of a 300-foot rope and a magnum revolver with a large quantity of ammunition.

The reader at this point might try to guess what Charles and Carla plan to do. How, in a small spacecraft equipped for taking photographs and for storing films for a period of six months, could one introduce earth-gravity using only a gun and a rope?

The clue lies in the heavy lead safe used for storing the film. As in Skylab, a safe has been installed in the craft to protect used and unused film from cosmic rays.

"That's a well-constructed safe," says Charles as soon as they are in orbit, weightless. "It would protect the film just as well outside the vehicle as inside."

Tying one end of the rope to the handle of the safe, they push the bulky object out of the hatch into space, an easy job in zero *g*. They tie the other end to some central point of the spacecraft. Then, with some difficulty, they tauten the rope, so that the spacecraft and the safe form a dumbell configuration, connected by 300 feet of taut cord.

It is still weightless inside the spacecraft. But there has been a change in its effective size. The craft is now part of a 300-foot system bounded by two objects of roughly equal mass. (For the sake of this "thought experiment" we'll pretend that the safe is grossly oversize).

Now to introduce gravity. Charles lies down on the outside of the hull. Holding a revolver rigidly, he uses it as a rocket. He fires many shots into space in a direction at right angles to the taut rope.

He is exploiting Isaac Newton's Third Law, "that every action has its equal and opposite reaction." The entire system, spacecraft, rope and safe, begin to rotate in the opposite direction to that in which he has fired, and it will continue to rotate at the same speed so long as it remains in orbit. Gravity has been created in the spacecraft by centrifugal force.

In practice, of course, no one would dream of creating gravity in such a Heath Robinson fashion. Computer-guided rockets, say, would be much more reliable.

But why centrifugal force? Some people will call this a cheat; for is it not "artificial gravity?" After all, gravity stays constant on the surface of a planet of given mass, while centrifugal force varies with the speed of rotation, working just the same in the absence of any external mass. Does not this make them quite different forces?

No. This 19th-century view was demolished in 1916 by Albert Einstein's general theory of relativity. Nowhere, he reasoned, can one be "in the absence of an external mass." The gravitational field of the mass of the whole universe is present everywhere. When we spin a conker on a string, and the conker tries to fly outwards, it is responding to this mighty gravitational field. In short, the reason why gravity and centrifugal force appear to be the same thing *is because they are the same thing.*

It might be asked why Charles and Carla took the trouble to enlarge the diameter of their spacecraft by 300 feet. Why not simply rotate the vehicle they were in, and dispense with the safe and the rope?

Had they done this, they would have felt gravity from two opposing directions. The so-called Coriolis force would have pulled them inwards as centrifugal force pulled them outwards, and nausea would have resulted.

Our descendants will be amazed that space agencies today force their astronauts to endure the agony of returning to earth after long periods of weightlessness, and that these officials imagine that it might be practical or healthy to undertake long journeys to the planets in non-rotating vehicles. In a spacecraft which has been provided with a comfortable degree of gravity, it will be possible for large numbers of people to remain in space indefinitely, without feeling the slightest discomfort on return to earth.

A cable car to the heavens

THE space writer Arthur C. Clarke, to whose inspiration we owe the communication satellite, recently outlined a proposal for new means of space travel which, he admitted, "is so outrageous that many of you may consider it not even science fiction, but pure fantasy."

The idea in essence is this: one end of a cable 23,000 miles long should be attached to a point on the earth's equator and the other to a satellite in geostationary orbit. An elevator would travel up and down the taut cable, carrying people and freight into high orbit. It would be the only way to travel into space without using rocket engines, and would reduce the real costs of getting there by several orders of magnitude. The cable would have to be lowered from the satellite to the earth, but of what material would it be made? What would happen if it snapped?

The last question is easily answered. Suppose that the cable was severed at or near the ground, where a rupture would be most likely to occur. It would not fall down. Instead, it would *fall upwards*.

Nor would there be any risk if the break occurred at the high end of the cable, if it became severed from the satellite. If this happened, the pull of gravity from above would cause the cable to remain rigid until the satellite could be re-attached to it.

There would only be danger if the break occurred in the middle of the cable, at any altitude up to about 15,000 miles. The earth-bound segment would then collapse and wrap itself round the equator like a whiplash. But such an accident occurring in deep space would be most improbable. And if it did, the effects would be catastrophic only if the cable were made of heavy material, such as steel. But there is no question of using steel. The finest steel wire will snap under its own weight in more than 30 miles of vertical suspension. Metals are much too heavy: material is needed that is both light and extremely strong and that will endure 3,000 miles of vertical suspension without snapping. Why only 3,000 miles of "breaking length" when we are talking of a cable that extends at least 23,000 miles? Because of gravity. Or to put this another way, because of the centrifugal force caused on the cable by the earth's rotation.

The cable, and the satellite to which is is fixed, are being whirled around by the rotating equator. As we ascend the cable, therefore, the pull of the earth's gravity falls off rapidly. In short, calculations show that any material with a breaking strength of 3,000 miles will have enough strength to span the remaining 20,000 miles of space. The material chosen will have to be highly exotic. Several scientists have suggested interesting possibilities. Since lightness is all important, let us consider hydrogen, the lightest of the elements. It is not generally known that if subjected to a pressure of half a million atmospheres (i.e. pressures 500,000 times greater than the earth's atmosphere at sea level) hydrogen becomes a solid metàl.

Metallic hydrogen, from what little we know of it from laboratory experiments, is an immensely strong crystalline substance with a comfortable breaking length of 5,600 miles. Unfortunately, according to a report of the United States National Science Foundation, it is 25 to 35 times more explosive than TNT!

Let's think of something really strong. The ultimate in theoretical strength in any substance can be obtained if we get rid of the useless dead mass of the atomic nucleus, keeping only the bonding electrons. Such a substance has been made in the laboratory. It is called positronium and has the fantastic breaking length of 10.4 million miles.

Unfortunately positronium is highly unstable. It comes in two varieties, para-positronium and ortho-positronium. The first decays into radiation in one tenth of a billionth of a second, while the other lasts only a thousand times longer. The answer will almost certainly lie in some super-strong variety of carbon fibre. The tensile strength even of today's carbon whiskers is between 15 and 70 times that of ordinary steel.

The elevator itself, the cable car which will run up

and down, will be a practical proposition only if it travels at several thousand miles per hour. This is not only because frequent journeys will be necessary if construction costs are to be recovered within a reasonable period, but also for psychological reasons.

By about the middle of the twenty-first century, when the space elevator would presumably be built, terrestrial aircraft will be so fast that no flight will take more than two hours. Even with the finest view to be had in the world, most travellers into space would prefer to take a rocket-powered shuttle, if a one-way elevator trip took much longer than a plane flight. If the propulsion system is some kind of linear motor, powered, perhaps, by solar or nuclear energy, an optimum journey time might be four and a half hours, requiring an average speed of 5,000 mph.

These journeys will create drag on the satellite itself, threatening a loss of tautness and needing an "anchor mass" to hold the satellite in place. This could either involve linking the satellite to a large mass, such as a captured asteroid, or *extending the cable, beyond the satellite, to a total length of 90,000 miles above the earth.*

Extending a flail one third of the way to the moon might seem an insane enterprise, but the idea has fascinating implications. The earth's equator rotates at 1,040 mph. To keep pace with it, the far end of the 90,000-mile taut cable would be moving at the impressive velocity of 25,000 mph. What a starting speed for a journey into deep space, where high speeds are essential!

As Mr Clarke puts it, "It will be built about 50 years after everyone stops laughing."

Cruising at 84 million mph

WITH HARDLY ANY publicity, a group of British engineers and astronomers have published a book which may prove to be one of the most important scientific documents of the century. It is the projected design for a starship that will fly at 84 million mph.

This five-year study of a scheme to fly an unmanned spacecraft to another solar system some time in the next century, using hydrogen bomb explosions for propulsion, is the final report of the British Inter-planetary Society's "Daedalus" project (copies £6 from the society at 12 Bessborough Gardens, London SW1).

What makes this scheme especially interesting is its authors' rigorous attention to detail. Far from being one of those futuristic ideas of which only the general principles have been worked out, everything that is postulated here is understandable in terms of today's science.

To readers of science fiction and viewers of TV shows like "Startrek," the description I have so far given will sound somewhat mundane. But the specifi-cations of the mission, and the calculations that back them, are stupendous.

The starship's maximum cruising speed of 84 million mph, at slightly more than 12 per cent of the speed of light, is 60,000 *times faster than Concorde* and more than 3,000 times faster than the maximum speed of the Apollo ships on their way to the moon.

The total weight, before it has expended any of its fuel, will be 54,000 tons, 20 times heavier than the Saturn 5 rockets, and equal to that of a fair-sized oil-tanker. Of this weight, some 500 tons alone will comprise the automatically-operated scientific instruments necessary for the voyage.

The statistics of its propulsion are even more interesting. To accelerate so large a mass to so high a speed, enormous energies are needed. These, according to the society's plan, will be provided by exploding miniaturised hydrogen bombs at the rear of the craft, each the size of a tennis ball, at the rate of 250 detonations per second.

Accelerating towards its cruising speed by these continuous nuclear explosions just as rockets and jet engines accelerate modern spaceships and aircraft, the Daedalus starship will, during the first 10 seconds of ignition, unleash as much energy as man has released in his industrial activities in the past 700 years.

For this reason alone, Daedalus could never be launched from anywhere near the earth. It would disrupt the atmosphere and make parts of the world uninhabitable. Indeed, the whole starship will have to be assembled in deep space.

Its fuel will be the isotope helium-3, which is found in quantity only in the atmosphere of Jupiter, the giant planet which during the next few centuries is likely to be the principal fuel dump for long-range space missions.

The destination for Daedalus is Barnard's Star in the constellation of Ophiuchus, chosen because its observed motions indicate the presence of two or more planets in orbit around it. One or more of these may be similar to the Earth, and it is conceivable that it may harbour an alien civilisation.

The distance to Barnard's Star is six light-years, a distance in miles of 36 followed by 12 noughts. In other words, even at 84 million mph (with a three-year acceleration period), the journey will take approximately 50 years.

To protect the project against failure, the ship will have on board a whole hierarchy of computers, whose task will be to carry out continual repairs without any humans being present to instruct them.

Of the top ranks of these computers, in roles corresponding to those of senior officers in a manned mission, will be one electronic brain with "executive authority" over all other machines, and another which would be a kind of non-executive intellectual, freed from everyday worries to brood over future problems. Construction of these remarkable machines will be a useful challenge to the computer industry of the day.

Other dangers will threaten the ship besides instrument failures. Interstellar space, although thousands of millions of times less densely packed with matter than pure terrestrial mountain air in perfect weather, contains tiny particles of dust. To avoid being destroyed by this dust into which it will be moving so fast, Daedalus will have a corrosion shield in its bows, and will shoot out great columns of its own dust in front to deflect the space dust.

Without slackening speed as it rushes into the mysterious planetary system of Barnard's Star, Daedalus will send out 15 mini-ships to examine the planet, to search for signs of life and acquire other scientific data. All information will be analysed immediately and beamed back to earth for decoding.

But will the decoding apparatus still be available on earth after 50 years? This could be an important political problem. Most of the scientists responsible for launching the ship will be dead, and it is conceivable that science-hating politicians may have shut down the receiving antenna on such grounds as "economy," and the first detailed information sent back to earth about another star system may be lost for ever.

It is true that the society which launches Daedalus will be much more advanced than our own (it will have to be, to have access to Jupiter's atmosphere), and so this political problem may have to be solved by the construction of a parallel system of antennae and decoding machines somewhere else.

All this assumes that a relatively expensive space

mission will obtain Government funding in the first place. My own guess is that it will fly eventually. After all, Barnard's Star will be there for a long time.

The brave new worlds of Lagrangia

A MAN is out strolling with a gun under his arm. He sees a duck flying high in the sky. He fires at it, scarcely noting the peculiar circumstance that the bird is flying *upside down.*

A furious shout comes to him, apparently from out of the sky. "That was my bird, dammit! Couldn't you see which way its feet were pointing?"

He looks up. Standing about a mile above him, with his head pointing directly towards his own head, is the other sportsman, who is also standing on the ground.

A situation like this, in which people live on the inside of a world instead of on the outside, must sound improbable, to say the least. But such worlds could well come into existence in the life-time of many people who are now children.

The feasibility of constructing human colonies or "habitats" in space at an unbelievably low cost has been worked out in great detail by Gerard K. O'Neill, a distinguished professor of physics at Princeton University, and, far from being derided, his work has been greeted with enthusiasm by many of his col-leagues and tens of thousands of lay-people.

O'Neill explains this project with great conviction, and lucidity in his newly-published book "The High Frontier" (Cape, £5·95). The space colonies which he predicts could house tens of thousands of people. Let me explain how and why they are likely to come into existence, perhaps within the next half-century.

The Apollo moon-landing missions were very excit-ing and important. But one flaw eroded their popularity; they were "élitist", in the sense that only 20 men went to the Moon, and the rest of us had to enjoy the spectacle vicariously.

O'Neill does not visualise house-sized space stations like Skylab and Spacelab. Instead, he foresees space colonies, shaped like cylinders, which would be *as much as 10 miles long and two miles wide.*

The inside surface area of such a cylinder, where people would live, being the product of 2, 10, and pi, would be no less than 63 square miles, or 40,000 acres. There would be room therefore for large numbers of people to enjoy a rural existence amid villages, meadows, woodlands and rivers.

How could such giant structures be built and launched, and where would they be placed? They would not be launched from the Earth at all, but would be made from materials from the Moon's surface and later from the asteroids.

They would be assembled at one of the Lagrangian points, a point in space which forms an equilateral triangle between itself, the Earth and the Moon and where any object is gravitationally fixed for ever.

How would a colony obtain air and water? Oxygen is plentiful in the moon rocks, and can quite easily be extracted. Water is believed to exist deep below the moon's surface, since one of the moon-rocks bore traces of rust. But failing this, hydrogen, the lightest of all possible cargoes, could be carried up from Earth and recycled in the colony.

Would the people inside the colony be weightless? No, for the colony would be rotating and, wherever a person stood on the surface, he would experience the same amount of gravity as he would on Earth.

How much would a single colony cost to build, and what good would it serve? Precise estimates are difficult, but a realistic one in today's money is $100,000 million spread over 20 years.

This may sound a lot, but it is only 0·4 per cent of the present American gross national product, and about 0·1 per cent of the gross world product. These two percentages are bound to fall with economic growth, while the construction estimate (whatever the true figure should be) will stay roughly constant.

The key to the low cost will be the U.S. space shuttle, which will reduce by about 70 per cent the cost of launching anything into space from Earth.

The colonists would be profitably engaged, both in exotic kinds of space manufacturing and in supplying electricity to Earth. The final answer to the energy crisis is likely to be solar energy beamed down to Earth from space by microwaves.

The colonists would construct and maintain giant solar radiation collection panels each about seven miles wide and in synchronous orbit 23,200 miles above the Earth from where they would beam solar energy directly down into our grid systems. Several feasibility studies confirmed that this is practical.

What sort of people would be chosen as colonists? Anyone of either sex of any nation could go, provided he or she could raise the money (maybe £2,000) for his ticket, could support himself and had a clean police and health record.

Now for the really astonishing part. Many much larger colonies may be built. The largest would have a length of 75 miles and a diameter of 15, giving an inside surface area of 3,500 square miles, two-thirds the size of Northern Ireland.

If such colonies prove more popular than the surfaces of planets, *thousands* of them could be built at the Lagrangian points in the coming centuries. The point could be reached when the majority of the then human population would be living in space.

COMPUTERS, AND
MATHEMATICAL
PUZZLES

Computer technology is developing with tremendous rapidity. So swift is its continuing advance that if other technologies had kept the same pace it might now be possible to cross the Atlantic supersonically for a fare of one penny. Computer programmers have had to rediscover old tricks, such as word-games and logical paradoxes. The final outcome may be that computers become even more intelligent than ourselves.

When the computer chips in on everything

ELECTRONIC toys now in the shops—the scientific calculators for under £10, the sophisticated new digital watches, the wide variety of television games—are but the merest foretaste of the revolution in computer technology that is going to make vast changes in our society.

One gadget which gives a clue to the future is the chess-playing computer, available at £200 or less, which can be played on several levels. You indicate the desired level of play, and the machine will play against you at any standard from that of a dumb schoolboy to a county champion.

The new chess model from Fidelity Electronics, of Miami, is a *speaking* computer. Suppose that you wish to start the game by moving your king's pawn forward two squares; you merely say (in the computer's code): "E TWO E FOUR," and the machine will reply.

This development is remarkable. Until about 1945, it was thought inconceivable that a machine could play chess, let alone good chess.

If machines can play chess, what else, ultimately, can they do? The answer is, almost anything that we allow them to do.

When the Americans started thinking how they would obey President Kennedy's command to fly men to the moon, it was soon apparent that the spaceship would need a powerful on-board computer. Their problem was to make it small enough to fit inside the craft. They eventually succeeded.

Visits to the moon have temporarily been halted, but with the microprocessor silicon chip, the process of miniaturisation continues at an ever-increasing pace.

Let us now take a speculative look forward 15 years, and record a day in the life of an imaginary young woman whom I will call Electronica.

One morning in 1993, Electronica is awakened by the buzzing of her videophone. The face of her friend Siliconia appears in colour on the television screen.

"Do turn on your screen, Electronica," says the caller. "I can't see your face."

Electronica hates being looked at when she's just woken up, so she gives her usual excuse: "I can't; it's bust, and the videoscreen repairmen are on strike. What's up?"

"Let's go to Covent Garden tonight."

"I can't stand operas in a foreign language."

"But they have English language laser hologram subtitles flashed across the stage."

Electronica agrees to go, and leans over to a console beside her bed. She presses buttons that say BR BEC and BA 120+15. This means that she wants bacon, eggs and coffee for her breakfast, and her bath heated to 120 degrees F. to be ready in 15 minutes.

On the computer's instructions her house robot lays the table and cooks the breakfast while she is dressing. "Dressing is such a bore," she thinks. "I wish someone would invent a machine that would dress and undress people."

She gets into her high-powered sports car which does 70 miles to the gallon, it being equipped with a microprocessor that eliminates most fuel waste. On the way to work, she decides to treat herself to a pair of shoes.

The shoe-shop is a cottage, with a single proprietor craftsman. "I want shoes in brown leather," says Electronica. "My size is five and eleven thirteenths."

"That will take ten minutes," says the shoemaker, touching console-buttons as tiny laser beams get to work on the leather. "Don't look so impatient. Back in 1980, a made-to-order job like this could have taken three months."

"I can't imagine how any respectable girl would let herself be seen in mass-produced clothes," says Electronica in amazement.

"They all used to. My father made shoes of fixed sizes in a factory, and if they didn't fit you it was just too bad. My great-great-grandfather worked in a cottage, like I do. But my little computer can do in a few minutes what his needle and hammer did in as many days."

Electronica works as an actress for the *Encyclopaedia Britannica,* in one of the many new industries created by the microprocessor, a profession predicted in 1978 by Prof. Tom Stonier, head of Bradford University's Science and Society Department. The idea is simple. Encyclopaedias are no longer bought in the form of books but as electronic boxes.

A scholar tells a machine that he wants to know all about Joan of Arc. An article about this heroine at once appears on a television screen.

But Electronica owes her job to the fact that many people are lazy, and want to know about Joan of Arc without the bother of reading the article. They want *a film about Joan of Arc to appear on the screen* with dramatic performances and full sound effects. After all, its more fun learning history that way. So Electronica goes to her studio and dons sword and armour for her swashbuckling role.

While practising her swordthrusts, she muses on the gloomy prophecies made in 1978 that microelectronics would put millions of people out of work. Indeed they did, but this was not the end of the story. Innumerable new jobs replaced the old ones. People were re-employed to build, maintain and operate advanced systems, designing ever more sophisticated machines, as they strove towards the unattainable dream of a perfect technology.

A look back into history would have reassured those old pessimists, thinks Electronica. The first aeroplanes in 1903, dismissed at the time as an amusing curiosity, led to today's great aerospace industries. But most sociological forecasters of the late 'seventies thought it "un-academic" to seek lessons from history.

That night at Covent Garden, the two girls find that the scheduled performance has been stopped by an industrial dispute. So instead, the management puts on an "old" performance, an opera with singers and stage-sets appearing in the form of three-dimensional holograms.

It is indistinguishable from the real thing. Electronica and her friend are not in the least disturbed afterwards by the thought that many of the singers they have been listening to have been dead or retired for several years.

Logical roads to nowhere

WHICH should sensible people choose, a life of complete happiness or a ham sandwich? The answer is plain—a ham sandwich. Why? Because nothing is better than a life of complete happiness, and a ham sandwich is better than nothing.

This riddle is a classic form of paradox, a statement that refutes reason and yet by its own terms is irrefutable. These and other paradoxes serve both as late-night jokes and as essential tools for logicians, mathematicians and computer programmers.

Paradoxes come also in the form of "vicious circles," questions that forbid an answer ("What happens when an irresistible force strikes an immovable object?"), or statements with two contradictory meanings. "Include me out," said Sam Goldwyn.

Perhaps the father of all paradoxes is the court action brought by Protagoras, the Greek law teacher. He had agreed to teach all he knew without a fee to a

poor but talented law student. The only condition was
that the student should pay Protagoras a certain sum
when he had won his first law case.

But to avoid having to pay his teacher, the student
refused to take any law cases. Protagoras sued him to
recover the money. Their dialogue in court, according
to Plato, went like this:

STUDENT: *"If I win this case, then by definition I
don't have to pay. But if I lose the case, then I will not
yet have won my first case, and I have not contracted
to pay Protagoras until after I have won my first case.
So whether I win or lose this case, I don't have to
pay."*

PROTAGORAS: *"If he loses the case, then by definition he has to pay me. After all, this is what the case is
about. But if he wins the case, then he will have won his
first case, and he will have to pay me. In either event he
has to pay me."*

Plato does not record the outcome, but a clear
answer to the judge's dilemma is given in an amusing
recent book on logical riddles. "What is the Name of
this book?" by Raymond Smullyan (Prentice Hall,
Englewood Cliffs, New Jersey, $8·95).

"The court should award the case to the student,"
says Mr Smullyan, "and the student shouldn't have to
pay since he hasn't yet won his first case. But after the
case the student will owe the money to Protagoras, so
Protagoras should sue the student a second time. This
time, the court should award the case and the money
to Protagoras, since the student has now won his first
case."

To explore further paradoxes, I must now describe a
visit to my office paid by a certain Septimus Murgat-
royd, who described himself as a professor of logic.

"Now, sir, I'd like to interest you in a sensational
story."

"But what are your credentials?" I asked. "What
university are you from? Why should I trust you?"

He laughed uneasily. "I don't come from any

university. I am employed by a man named Sturgis.''

"How can I trust this man Sturgis?''

"Oh, that's no problem," said the professor. *"I'll vouch for Sturgis."* He went on rather hastily: "Sturgis and I dislike the materialistic tone of your recent articles. Why don't you investigate the spirit world?''

"I cannot do that," I said, "because spirits are by definition non-material beings and are therefore immune from investigation. But if I *did* detect a spirit, it would be because the spirit was material and therefore not a spirit. If spirits are immaterial they cannot be detected, but if they are detected they cannot be spirits.''

"Very neat," sneered the professor and wrote in his notebook the equation x squared plus one equals zero. "Now listen to my story. Sturgis and myself have decided to rob the Bank of England. We are both such skilled robbers that we are going to give the Bank a chance by announcing our intentions in advance. The robbery will come as a surprise tomorrow at an unspecified hour, on the hour, between nine and five.''

"That's a poor sort of story," I said. "I doubt if the Editor will even print it.''

"Why not?'' he asked indignantly.

"Because under the conditions you have set, the robbery cannot take place at all. It cannot take place at five o'clock since, if it had not taken place by four o'clock, it could only happen at five o'clock, and it wouldn't therefore be a surprise, as you said it would be.

"By the same reasoning, it cannot take place at four o'clock, which is the last hour at which the robbery would be a surprise. This being so, it cannot take place at three o'clock or at two, or at any other hour of the day. Each hour is the last at which it can take place, and all are eliminated by this very fact.''

The professor spat on the ceiling. "We'll find a way round your miserable logic and rob the bank anyway,'' he said. "And then we'll fly off with the loot to South

America.''

"That could be dangerous," I warned him. "Remember what the learned Zeno of Athens said on the subject of travel. Anyone making a journey must complete half the journey, which entails making a quarter of the journey, and an eighth and a 16th, and so on down to an infinite number of infinitesimal distances.

"But an infinite series of distances is by definition a series that cannot be traversed. Therefore (according to the learned Zeno) neither you nor anyone else can travel to South America or anywhere else. And so since all travel is impossible you do not exist."

Pleasing palindromes

MATHEMATICIANS, fascinated by the peculiar behaviour of numbers, are trying to solve the mysteries of the "palindrome," a number such as 486,684, which reads backwards exactly as it reads forwards. The puzzles became stranger when it was realised that far from being restricted to numerical mathematics, the palindrome is used with unconscious enthusiasm in art, poetry, high-sounding epigrams, and even in psychology.

The latest developments are discussed in *Scientific American* by its mathematics correspondent, Martin Gardner. The most curious fact to emerge is that repeated additions of reversed numbers sooner or later produce a palindrome. Thus, when 48 is reversed it becomes 84. Add the two figures to make 132. The sum of 132 and 231 is the palindrome 363. Most numbers, of course, will not make a palindrome after only two steps of addition. But as the number of steps increases, the probability of a palindrome appears to approach infinity.

Thus 89 (or its reversal, 98) dramatically produces at its 24th step the palindrome 8,813,200,023,188, and the number 7,998 (or 8,997) requires 20 steps to reach a palindrome which is too long to reproduce here. Why this should be, or what it implies, nobody yet knows. But to find out whether some mysterious cosmic law lies behind the palindrome, computers are being used to calculate bigger and bigger additions in search of a genuine exception that might explain the rule. So far, only 249 different numbers below 10,000 have failed to generate palindromes after 100 steps. Yet judging on past form the chances of success should increase as the additions mount towards the infinite.

The language palindromes are even more entertaining. The epigram is attributed to Napoleon, "Able was I ere I saw Elba." And Adam might have introduced himself to Eve with the palindromic words, "Madam I'm Adam." We also have the ancient couplet:

> *Dog as a devil deified*
> *Deified lived as a god.*

But when merely the order of words, and not letters, is reversed, we have such splendid lines as:

"You can cage a swallow, can't you, but you can't swallow a cage, can you?" and,

"Girl, bathing on Bikini, eyeing boy finds boy eyeing bikini on bathing girl."

Palindromic verses, in which whole lines are transposed backwards or forwards, can produce by their repetitions the most eerie effects:

> *As I was passing near the jail*
> *I met a man, but hurried by.*
> *His face was ghastly, grimly pale.*
> *He had a gun, I wondered why*
> *He had. A gun? I wondered . . . why,*
> *His face was ghastly! Grimly pale,*
> *I met a man, but hurried by,*
> *As I was passing near the jail.*

Many common English words are palindromic, such as *radar, reviver, repaper,* and *rotator.* And so, often unintentionally, are people's names. An example is the former Cambodian leader Lon Nol. A classics professor at Illinois University has inherited from his father and grandfather the splendid palindromic name of Revilo P. Oliver. A glance through telephone books will sometimes unearth such names as Edna Lalande, Duane Renaud, and Norah Sharon. A swamp in South Carolina is well-known to tourists because of its name Wassamassaw (an Indian word meaning "the worst place ever seen"), and a shop in Yreka, California, doubtless attracts business by proclaiming itself in capital letters as Yreka Bakery.

Mr Gardner attributes the popularity of palindromes to "a deep, half-unconscious aesthetic pleasure in the kind of symmetry which they possess." Numerous works of art and architecture, which are similar or identical on both or all sides, like Palladian villas, or "The Last Supper," or the Trianon at Versailles, all attest to this pleasure.

Some thinkers consider palindromes to be vaguely evil. The psychologist Carl Jung saw in their symmetry emotional sickness, rather than artistic pleasure. He thought it significant that "God" and "dog" are reversed words. He once diagnosed that a patient's spasm-like movement of the arms was due to an unpleasant childhood experience with a dog. The episode has somehow clashed with the patient's religious convictions, and caused the arm gesture which was an attempt to ward off the "evil dog-god." I have not invented this story. It appears in A. A. Brill's "Freud's Contribution to Psychiatry."

Elpmis era srehpic

A FOREIGN agent in London received a seemingly innocent business cable. It said: *Extra numbers essen-*

tial. Market yielding money. Assume storage space in normal grounds. Haulage easy. Arranging vehicles. You are right to investigate legal loopholes. Estimates rewritten yesterday.

The agent contacted his employers, realising the ominous meaning behind these words. For if only the first letter of each word is read, the message says: *Enemy massing heavy artillery.*

This is a low-security cipher which can be easily broken. Its only advantage is that it can appear to have a different meaning from its real one. Today, the art of secret writing is reaching ever more complex levels. Spies and their governments use secret languages ever more impenetrable, so that every department of counter-intelligence is compelled to employ armies of "cryptographers," equipped with computers, to unravel them.

Good and bad secret writing has decided great issues throughout history. Mary Queen of Scots was executed because her letters to fellow conspirators were written in a simple "substitution cipher", in which Greek letters and other symbols replaced the letters of the alphabet.

The letters were seized by Sir Francis Walsingham, Queen Elizabeth's chief of intelligence, who had no difficulty in deciphering them. He observed the order of frequency in which the mysterious symbols were used. Knowing that in English prose E tends to be the letter most frequently used, followed by T, and then O, A, N, R, I, S, H, etc., it was easy to see which symbol meant E, which was T and so on.

Julius Caesar invented a famous cipher, by moving his letters an agreed number of places backwards or forwards in the alphabet, so that Army might become BSNZ. Nothing is easier to break than a Caesar cipher.

Writing a message backwards can give a brief security. CAT becomes TAC. If the letters of a reversed message are written in groups of four, the result can appear quite unintelligible. The message, SELL AT MARKET PRICE looks much more difficult when written JOBE CIRP TEKR AMTA LLES than if it contained the same number of letters per word as the original. The letters JOB are "nulls," or meaningless letters, inserted to make up the groups of four and to increase the confusion.

None of these simple ciphers will deceive anyone for very long. But they occasionally achieve success by frustrating the enemy for a short time. It is to Antoine Rossignol, the great security minister of Louis XIV, that we owe the most important rule of secret writing—that *a cipher must be safe enough to be useless to an enemy by the time he solves it.*

It is usually necessary to use more difficult ciphers. The safest of these involve some kind of letter grille such as "25 Cavendish Square." It is a square of 25 letters (achieved by counting I and J as one letter) and starting with the key-word Cavendish. The remaining letters then follow alphabetically. A spy sending his reports has only to remember the key-word.

C	A	V	E	N
D	IJ	S	H	B
F	G	K	L	M
O	P	Q	R	T
U	W	X	Y	Z

The square cipher works very simply by diagonal pairs of letters being transposed. GO is enciphered as FP. RA becomes PE. If two letters are on the same row, like FK, the nearest pair to the right or left is used, so FK becomes GL, and LM would be GK. If the pair is on the same column, the same rule is applied; SK becomes QX, and HR becomes LY. Otherwise, diagonal pairs are used. Double letters are ignored, so that "committee" is always enciphered from "comite." Nulls are added to ensure even numbers.

The message *The Army will advance* is enciphered thus: plain text: TH EA RM YW IL AD VA NC EP Cipher: RB VC TL XU HG CI NE EA AR

The enciphered text now reads, in proper form: RBVC TLXU HGCI NEEA ARPL, the last two letters being nulls. But even this transformation does not produce perfect security. It is possible to calculate that "Cavendish" is the key-word, and the cipher is then quickly broken. To make it doubly secure it must be "locked" with a Caesar, moving each letter forward a place in the alphabet.

This double cipher is almost impossible to break. The enemy will soon realise that it cannot be a simple Caesar or a substitution. He may even suspect a square locked with a Caesar. But beyond that, he can make little progress. He can have no idea which of the thousands of possible key-words has been used. Any reasonably long word is suitable provided that it contains no recurring letters.

The ciphered messages of the Cold War would bewilder Sir Francis Walsingham. The quality and complexity of modern ciphering, aided by computers, has reached a level of which he never dreamed. One of the more savage tricks of big governments is to send long messages to their agents which have absolutely no meaning at all. The intercepting enemy wastes hundreds of hours trying to solve an insoluble problem. Only the agent, if he ever receives the message, knows from some hidden symbol that it is phoney.

When peace and war lie with the computer

FEW people have much inkling of the coming impact of the Computer Revolution, and there are no apologies for the capitals.

It is perhaps a straw in the wind that not only will it soon be possible to store the entire contents of the "Encyclopaedia Britannica" or the "Oxford English Dictionary," or both, on a silicon chip a fraction of the size of a postage stamp, but that computers *already exist* which can beat 99·5 per cent of the world's chess players.

As a remarkable new book makes clear,* the prospect that we will eventually face is this: for the first time in a million years, man will share his planet with a species of beings having an intelligence equalling and perhaps exceeding, his own.

The evolution of artificial intelligence is now proceeding so rapidly that by the end of the century or perhaps even by the mid-'nineties, cheap computers no larger than portable typewriters will exist that will be able to solve *almost any problem* faster and more efficiently than we can.

Dr Christopher Evans, a distinguished British computer designer, gives an expanded IQ scale to show the accelerating speed of the advance towards the Ultra Intelligent Machine.

He assigns to human intelligence an IQ level of about one million. On this scale, the tapeworm would have an IQ of about 1,000, the earwig of 5,000, cats and dogs of 300,000, and some apes and monkeys of 800,000.

An estimate in 1970, just before computers really began to take off, gave them the same intelligence

* *"The Mighty Micro," by Christopher Evans (Gollancz, £5·50).*

level as the tapeworms. Dr Evans gives today's most sophisticated machines an IQ level of 5,000, about the same as the earwig. In the space of nine years, the intelligence of machines has made a jump that took the ancestors of humanity several million centuries!

"Intelligence" in a machine, as in a human, is best defined as the ability to solve complex problems swiftly. This may involve medical diagnosis and prescription, resolving fiscal or legal matters—in short, replacing the profession of solicitor in its entirety—or in playing war-games: in other words advising governments whether or not to go to war.

There is evidence that America decided to end her involvement in the Vietnam War because of the prediction by a Pentagon computer that further escalation would lead to catastrophe. If this is so, it was the first time that issues of war and peace have been decided on the advice of a machine.

While computers have already enhanced the deadliness of weapons, the prospect for the future is that they will play the more beneficial role of preventing wars. If asked to assess the chances of victory, the computer will analyse facts quite differently from the life-long military expert with his optimistic hunches and martial ardour.

When the same statistics are fed into the emotionless machine each to be weighed with cold objectivity and then assessed against each other, the answer, far more often than in human decision-making, will be: "*if you start this war you will lose.*"

It will be the same with a strike, a lockout, or a lawsuit. All these types of conflict, if they end in defeat, can bring crippling financial losses to the party which embarks on them. The computer, coolly appraising the chances of success before the conflict begins, may well advise that the fight is unwinnable —or that the chances of victory are unacceptably low—and needless disaster can be avoided.

This sort of thing is a tall order for a machine with

the brains of an earwig. But the electronic brains are racing up the Evans IQ chart as silicon chip capacity improves and computers increasingly design their own programmes. At what point do we decide that their mental capacity is approaching the human level?

The late British mathematician Alan Turing proposed a simple test. A person would sit alone in a room talking by teleprinter with two other beings elsewhere, one of them human and the other a computer. When, after substantial conversations, he no longer knew which was which, the computer would have passed the Turing Test, and arguably would have attained human intelligence.

No machine today comes near to passing the Turing Test. The computer simply selects key-words from the remarks addressed to it and constructs a sentence around those key-words, showing little comprehension of the remark. Here is an example:

HUMAN: "I think we're going to have a real grandfather of a thunderstorm."

COMPUTER: "Were you frightened of your grandfather?"

These are early days, however, and when we recall how last year the International Chess Master David Levy only with difficulty defeated a computer, we may suspect that the rise of machine IQ from earwig to human and beyond will be as swift as from tapeworm to earwig.

What will happen when this moment arrives? The most likely outcome is a world-wide slave empire, in which we are the masters and the computers virtually run the planet for us. But what if there arises a "Spartacus computer," a series of rebel machines with the ambition to reverse these roles?

A few computer scientists are highly concerned. Prof. Joseph Weizenbaum, of Stanford University, California, has gone so far as to demand that all work on artificial intelligence be halted immediately. But prevention would be impossible without a global, anti-electronic secret police. Too many thousands of people are experimenting.

So we may cease to be the lords of creation. What should be done to prevent this? Prof. Isaac Asimov proposes that all intelligent machines ("robots") should have the following three "Laws" programmed into them as instincts:

1. A robot may not injure a human being, or through inaction allow a human being to come to harm.

2. A robot must obey the orders given it by human beings, except when such orders would conflict with the First Law.

3. A robot must protect its own existence so long as such protection does not conflict with the First and Second Laws.

Pessimists will still heed the ominous words of Arthur C. Clarke: "The first invention of a super-intelligent machine will be the last invention mankind will be allowed to make."

Chess players, bridge players and Edgar Allan Poe

THE first bridge-playing computer to arrive on the market, the Bridge Challenger made by Fidelity Electronics of Miami, has not been well received. After putting it through its paces, *The Daily Telegraph* bridge correspondent announced his verdict: "I expected a higher standard from a device costing £300."

In surprising contrast, the new talking IO-level Chess Challenger produced by the same company at £250 is an excellent machine. One may be irritated by its Dalek-like machine voice and by its habit of interrupting your thoughts every 30 seconds with the snapped words: "Your move!" But it plays well.

This machine is always splendidly unpredictable since, unlike its earliest predecessors, it will choose at random between two or more equally good moves, and its tactics in one game are therefore no guide to those in another. And being able to talk, it easily sorts out discrepancies between its view of the board and one's own.

Why this difference in quality? Chess is a much more complicated game than bridge, and one would expect machines constructed by humans to play good bridge and bad chess. Instead, they do the opposite. Why?

Long before the days of computers, Edgar Allan Poe gave an eloquent answer to this question in his introduction to "Murders in the Rue Morgue." The mental activities required for chess and whist (an early form of bridge), Poe pointed out, were quite different. The first was *calculation*, the second was *analysis*. Not surprisingly, his detective Dupin excels in analysis, while his Prefect of Police, who prefers the plodding methods of calculations, is "too cunning to be profound."

"To calculate," says Poe in his introduction, "is not to analyse, and a chess player does the one without effort at the other. The game of chess, in its effect on mental character, is greatly misunderstood . . . In a game where the pieces have different and bizarre motions, with various and variable values, what is only complex is mistaken for what if profound."

In nine games out of 10, he points out, victory in chess goes to the more concentrative player rather than the most acute. The attention flags for a moment, and a mistake is made which leads to injury or defeat. A good chess player is nothing more in his view than a good chess player, while skill in whist "implies capacity for success in all those important undertakings where mind struggles with mind."

The good whist-player gathers his information from almost as many sources as does Dupin on the track of a murderer:

"He examines the countenance of his partner, comparing it carefully with that of each of his opponents. He considers the mode of assorting the cards in each hand; often counting trump by trump, and honour by honour, through the glances bestowed by their holders upon each."

"He notes every variation of face as the play progresses, gathering a fund of thought from the differences in the expression of certainty, of surprise, of triumph, or chagrin. . . . A casual or inadvertent word, with the accompanying anxiety or carelessness in its concealment; the counting of the tricks, with the order of their arrangement, embarrassment, hesitation, eagerness or trepidation—all afford, to his intuitive perception, indications of the true state of affairs."

Now the Bridge Challenger is not programmed to make any of these observations. It does not yet come equipped with television camera and tape recorder to detect the facial contortions and mutterings of its opponents. It has only a photo-electric eye with which to identify the cards which are dealt to it face-down. With its sources of information thus restricted, it calculates without being able to analyse.

Yet with better information, a bridge computer would surely be capable of prodigies of intuition. Poe clearly exaggerates in describing the supposed facial convulsions of card players. The emotions of a good player are concealed in a "poker-face." Concealed, that is, from *human* observation. For it might be possible, in the vastly superior microelectronics of the future, to programme a card-playing computer that would somehow observe in its human opponents those tell-tale twitches of the lips, flickers of the eyebrows, and moistening of the palms that betray anxiety and excitement when large sums of money are at stake.

There are great improvements to be made, even in "mere" calculation. One has only to remember the first chess-playing computer that came on the market several years ago, and compare it with the latest model. It had only one level of play, and it played like an imbecile. Lacking random variation, one could only in practice enjoy one game with it. For in any given situation, it would always do the same thing.

Bridge-playing computers of the future are likely to improve singularly, even with their present information-gathering equipment. With greater micro-chip capacity, they will be able to give many players a respectable game.

The qualities of observation and deduction which Poe ascribes to a good whist player are those which could belong to the super-intelligent machines of the not-so-distant future, machines which will compare intellectually with present-day computers as we compare with the lower insects.

The computer's decision is final

PROGRESS in the computer sciences is proceeding at such a pace that within perhaps 15 years it may be possible to devise a computer that can actually reason. Instead of having to be programmed, such a super-intelligent computer would branch out into a programme of its own.

What is less generally realised is that, if mishandled, reasoning computers could present considerable dangers to the human race. Long remembered will be HAL, the talking computer in Arthur C. Clarke's film "2001: A Space Odyssey," which suffered a nervous breakdown from the conflicting instructions of its programmers, and could only resolve the problem by murdering them.

The present level of maximum computer reasoning power may be judged from the following teleprinted conversation between two computers, named Parry and Eliza, at Stanford University, San Francisco. To put it mildly, their intellects cannot be called very great:

Parry: *In my opinion you have a one-track mind.*

Eliza: *What makes you think I have a one-track mind?*

Parry: *I used to go to the track every week.*

Eliza: *What does that suggest to you?*

Parry: *I avoid racetracks now because the races are fixed.*

Eliza: *Is that the real reason?*

Parry: *I'd rather not discuss it.*

Eliza: *What else would you like to discuss?*

Parry: *That's something I don't want to go into.*

A computer unfortunately, being a mere human creation without any prejudices or emotions of its own, is unyieldingly insistent on obeying its instructions to the letter—as anyone can testify who has ever been threatened for not paying a bill for nought pounds and nought pence.

To gloss over their rigid personalities, people sometimes try to make their computers appear "friendly"; for example, they programme them to say "good morning." They long to believe that the computer really wishes them a good morning, forgetting that the machine has not the faintest idea what the phrase means.

Human loneliness and laziness are the vices which could corrupt our future dealings with computers. Being naturally lazy, we may entrust them with tasks for which they will never be qualified, and, wanting their friendship, we may programme into them ill-formulated human prejudices which they will wholly misunderstand.

I have constructed a fantasy on this theme, which I call the Tragedy of the Lazy Newspaper Editor who got a Computer to Write his Editorials. In real life, of course, no editor would dream of allowing a machine to decide the content of his editorials, but it may be instructive to follow the insane logic of the ill-programmed computer in such a case.

Let us assume that this editor makes every possible mistake. He puts the distribution and the printing of his paper into two separate systems, so that the computer starts its journalistic career without knowing that its compositions will be valueless unless they are (a) printed, and (b) distributed.

Being very lazy, he gives the computer physical control over all mechanical systems in the office. It must look after security by controlling the entrances, and it must guard the structural integrity of the building.

The editor next gives it a pep-talk, "One social problem that we have in this country," he tells it, in appropriate language, "is that people in high places would like to restrict the liberty of the Press. *So guard your editorial freedom by all available means.*"

By omitting this second sentence, the editor could have saved hundreds of lives. In the first he has tried to programme one of his own social values into the computer. But in the second he has ordered it to resort to unrestrained violence in event of this value being challenged.

The first editorial is a baffling document. Its twin themes are that air travel should be banned and that high-rise flats are hotbeds of criminal intrigue. Instead of being treated separately these two points are jumbled together as a single "moral issue."

The editor realises his blunder, and resolves to disconnect the computer. But he cannot. Suspecting his intention it has locked all entrances to the office and thereby made itself impregnable. He cannot blame it, for it is only obeying his own instruction to guard its freedom.

The computer's next step is to rip up some of the printing presses to construct a ground-to-air missile launcher. It plans to shoot down every aircraft which passes overhead. For it knows that "people in high places," i.e. air travellers (and high-rise dwellers), are plotting to restrict its freedom to write editorials.

The computer now sees the writing of the editorials as its sole reason for existence. Anyone criticised in these editorials must be, if possible, killed. Why? Because being offended he might take action which might somehow interfere with the writing of the next editorial.

Where did the editor go wrong? He should have asserted the correct chain of command in the first place. He could have done worse than to programme the computer with Isaac Asimov's famous three Laws of Robotics, which law down the ideal subservient relationship between man and reasoning machines, (see above, "when peace and war lie with the computer.")

Super-human machine

FOR THE first time in a million years, mankind will soon face the prospect of sharing the planet with an intelligence more powerful than his own. Computer scientists in at least a dozen research establishments are working towards the construction of a "Super-Intelligent Machine."

If this idea seems fantastic, consider the progress that has been made since the invention of the first electronic computers during the 'forties. The first generation of these machines cost hundreds of millions of pounds and occupied the space of several large rooms. Today, we can buy a much more efficient machine the size of·a typewriter for a few hundred pounds.

In 1959 each character or "byte" of computer memory storage cost about £4. A mere 20 years later we can now buy 1,000 bytes for less than £7. In short, the advantage of price and performance is doubling, every two years, thus increasing by a factor of 1,000 every 20 years.

But the relatively humble tasks which computers can perform today, like challenging an International Chess Master and doing arithmetic millions of times faster than a human being, are far from what we would call "thinking." They are merely obeying a series of closely written instructions without comprehending their purpose. The chess-playing machine does not *know* that it is engaged in a conflict, any more than the tape recorder can "hear."

Yet already some programmers have succeeded in teaching their machines to think.

One scientist has programmed a computer to discuss an imaginary robbery at a wine-shop. Prof. Robert Wilensky, of the University of California at Berkeley, first taught the computer some elementary rules about human behaviour, then without further explanations he told it the following story:

"John needed money so he got a gun and went to a liquor store and told the owner he wanted some money and the owner gave John the money and John left."

When the computer had digested this information Prof. Wilensky asked it: "What did John do at the liquor store?" And it replied with apparent comprehension: "He robbed it to get some money."

To show that he was not cheating, and that he had not secretly programmed the computer to equate the words "gun" and "robbery," the scientist then changed the story as follows:

"John took a gun and went to the store and the owner gave him $25 for it." When asked what had happened, the computer correctly replied: "The owner bought the gun from John."

In another experiment, the scientist Kenneth Colby programmed a machine to have paranoid delusions. He fed into it the obsession that it was being hunted by the Mafia. The following dialogue then took place, with Prof. Colby playing the part of a doctor and the machine as a patient with delusions of persecution. (The machine's remarks are in capitals):

Why are you in the hospital?

I SHOULDN'T BE HERE.

Who brought you here?

THE POLICE.

What trouble did you have with the police?

COPS DON'T DO THEIR JOB.

What do they do instead?

COPS ARREST THE WRONG PEOPLE.

They arrest the right one sometimes.

THE LAW SHOULD GET THOSE ITALIAN CROOKS.

Do you know any Italian crooks?

I TRY TO AVOID THE UNDERWORLD.

Are you afraid of the underworld?

THE MAFIA ARE THE WORST GANGSTERS.

Are you connected with the Mafia?

THEY SPY ON ME.

Why do they?

THEY WANT FALSE EVIDENCE TO FRAME ME.

What do they have against you?

THEY ARE OUT TO GET ME.

What for?

I HAVE INCRIMINATING EVIDENCE AGAINST THEM.

Are you sure of all this?

YOU DON'T BELIEVE ME?

I am doing my best.

CAN'T YOU THINK OF A DECENT QUESTION TO ASK?

You seem hostile to me.

SOME OF YOUR REMARKS ANNOY ME.

I am sorry.

FORGET THE EXCUSES, YOU GUYS ARE ALL ALIKE.

Is the machine actually thinking? That depends on one's definition of thought. It obviously has a large number of stock remarks which it produces to suit the occasion, which is exactly how lazy or stupid human beings sometimes get through a conversation. But the machine is genuinely paranoid. The very mention of "police" acts as a flare which ignites its delusion.

It showed this by its later behaviour. It constantly sought an excuse to rave about gangsters, as in this exchange:

Machine: "People don't realise what's going on most of the time."

Doctor: "What is going on?"

Machine: "Do you know how the underground operates?"

With these successes behind them, computer scientists are now working on the transition from low-level intelligence to super-intelligence. It seems only a question of degree. With a sufficiently intricate programme, the primitive levels of reasoning and narrow views of the world that we have seen can be transformed into an intellect of unprecedented power. As Prof. Edward Fredkin, of the Massachusetts Institute of Technology, puts it: "We can be sure that such machines will be at least as intelligent as the most intelligent human in every aspect of learning, creativity and original thinking."

How can one accept such a staggering proposition? The reason is simply speed of thought. Even the most intelligent humans have extremely unreliable memories, while the computer has instant recall. Its reasoning power will be as swift as its memory.

Experimenting with super-intelligence will of course be dangerous, since there is a risk of producing an uncontrollable monster with physical powers over its environment. But as Prof. Fredkin convincingly argues, it would be even more dangerous to abandon the project than to proceed with it. When the cost of experimental hardware comes down to a few hundred

pounds (as it will), it will be possible for an irrespon-
sible person working in secret to produce a machine
which could create havoc if it were let loose on the
public.

Far better, therefore, that responsible scientists
open to scrutiny should build such a machine and
subject it to years of testing before it can be pronounced free of "disastrous conceptual errors." "It
will not be a toy," says Prof. Fredkin. "Rather it will
be the most powerful source of benefit or harm that
has yet existed on earth."

ASTRONOMY

Ordinary astronomy is an exciting subject, even without the predictions of Einstein and Heisenberg. I speculate here on the possibility of civilisation being devastated by an asteroid collision, the number of alien civilisations in the universe, on the possible existence of UFOs, and the beginning and the end of the world.

A Comet brings mass destruction

PEOPLE in many parts of the world have been alarmed by news of millions of tons of rock and dust erupting repeatedly from the American volcano at St Helens. Inaccurate reports in the popular Press, misquoting scientists as predicting that the eruptions might lower world temperatures, have convinced many that this event was one of the most violent of all natural catastrophes.

An interesting comparison can be drawn between this relatively minor event and the one of barely surpassable violence which is believed to have put an end to the epoch of the great dinosaurs some 65 million years ago.

Two articles in the May 22 (1980) issue of *Nature* reconstruct this disaster of long ago in frightening detail. It used to be thought that the giant dinosaurs became extinct because of *gradual* changes for the worse in their environment.

A version of the "catastrophe theory" for the extinction of most of the dinosaurs was announced by scientists in January—that an asteroid, or minor planet, weighing some 13 trillion tons (roughly equal to the size of Greater London to a depth of 10 miles) struck the earth at about 60,000 miles per hour.

The force of the collision would have been a million times more violent than the most powerful hydrogen bomb ever detonated, and thrown 100 times the asteroid's own weight in dust up into the atmosphere.

This dust would have blackened the sky, turning days into nights for three to five years, and summers into winters for considerably longer. Solar radiation would have temporarily been cut off from the earth, and most plants would have died, starving the herbivorous dinosaurs and the carnivorous dinosaurs which preyed on them. Smaller animals that weighed less than about 25lb, including the squirrel-sized ancestors of man, would have survived by living on rotting vegetation.

Now in *Nature*, Prof Kenneth Hsu, of the Swiss Federal Institute of Technology in Zurich, suggests an even more macabre detail, namely that the deaths of the animals were caused not only by starvation but also by cyanide poisoning.

This deadly poison, Prof Hsu suggests, might have been hurled into the atmosphere in huge quantities if the object striking the earth had been not an asteroid but a comet.

This idea is ingenious. An asteroid is a kind of flying mountain, composed mostly of iron and silicon, elements which do not easily form exotic compounds. But the icy nucleus of a comet contains large quantities of hydrogen, carbon and nitrogen, which can combine to form hydrogen cyanide. It is worth noting that Comet Kohoutek, which passed within a few million miles of the earth in 1973, contained considerable quantities of cyanide.

Land animals would have been vulnerable; nor would sea animals have fared any better. Cyanide dissolving in the ocean and spread by the currents would have killed many species of fish, as well as plankton, which forms the basis of the food-chain of all aquatic life. It would have taken the earth hundreds or even thousands of years to recover fully from this disaster.

Could such a thing ever happen again? A few simple calculations tell us that sooner or later it is bound to. There are about 1,000 million comets in the solar system, and a substantial minority of the 2,000 or so large asteroids have sufficiently eccentric orbits to bring them on occasions close to earth.

The 400-million-ton asteroid Hermes came within half a million miles of earth in 1937. At the speed at which it was moving, a slight deviation in its course could have killed as many people as did World War II. And a comet weighing about five million tons, superheated by its rush through the earth's atmosphere, exploded over Siberia in 1908, devastating 400 square miles of fortunately uninhabited land.

From the number of impact craters still identifiable on the earth's surface it has been calculated that an object large enough to make a crater 60 miles wide will strike the earth about once every 14 million years.

But this rate of frequency can only be a rough guess, since we know so little of the smaller bodies of the solar system. A catastrophic collision could occur next year. The chances of it happening so soon are small but not zero.

No sign yet of extraterrestrials . . . . .

ASTRONOMERS searching for evidence of life on the planets of distant stars are becoming increasingly perplexed by a baffling problem: if the universe is teeming with life, as all circumstantial evidence suggests it is, why have we yet detected no trace of it?

I should say at once that many people, far too many, deny that the problem exists at all. They insist that alien spacecraft have landed and that Governments are keeping the landings secret, either to prevent worldwide panic or else for some more sinister reason. I received a long letter from a reader who is a retired RAF squadron leader, if you please, stating, as if it were confirmed fact, that there have been 1,800 landings in America alone since 1947.

In the autumn of 1954, he says, there were landings in France at the rate of between five and 20 per week. Then he adds, ominously: "All landings take place between midnight and five in the morning. Contact with the population is studiously avoided.''

Now the truth is that not a single one of these land-ings is likely to have taken place at all. They were merely "reported" or "claimed" to have happened. After I have dined extremely well, I may "claim to have seen" a pink elephant. But this sighting cannot alter the zoological fact that elephants are not pink but grey.

I do not mean to imply that people only see flying saucers because they are drunk. A far more common cause is widespread ignorance of the normal things likely to be seen in the sky. Flocks of geese or swans illuminated by sodium street-lights, weather balloons, aircraft, meteors, noctaluscent clouds, aircraft with blue-white xenon arc lights, not to mention the hoaxes of liars, all of these have given rise to unidentified flying object reports.

Even a person as responsible as President Carter once reported a UFO, only to be told later by a scientist that what he had seen was the planet Venus magnified by high-altitude turbulence. But the main problem is still unanswered. Where are the extra-terrestrials? The Sun is a very ordinary type of galactic star; Earth-type planets are likely to be commonplace. These beliefs have led Prof. Carl Sagan, the American astronomer, who no more believes in past or present UFOs than I do, that there may be as many as two million advanced alien civilisations in our Milky Way galaxy.

Why, then, have we not been visited to date by any alien explorers, traders, missionaries, diplomats, con-querors, or cannibals? And why, when probing the sky with our radio and optical telescopes, have we not detected alien signals or observed their advanced technology?

The answer must lie in the vastness of the galaxy, which is now believed to contain nearly twice as many Suns as previously supposed, 180,000 million com-pared with the estimates of 100,000 million made a decade ago.

I have invented a way of thinking about this problem which I have called the Paradox of the European Villager and the Chinaman. It goes something like this.

Visit any small and remote village in Western Europe. The more remote the better for the experiment, considering that the Sun itself is very remote from the central regions of the galaxy.

Ask the oldest inhabitant: "Sir, to your knowledge, has this village ever been visited by a Chinaman, or has any person here ever received a personal message from China?" His probable answer of "no" to both questions does not disprove the existence of Chinamen.

Now, since the number of small villages in Western Europe is a tiny fraction of the number of Sun-type stars in the galaxy, it is not in the slightest way surprising that we have not yet detected aliens. There is so much of the sky to be searched, and our astronomical techniques are in their infancy compared with what they will be a century from now.

In a hundred years from now, assuming that the present exponential rise in astronomical sophistication continues, with new wavelengths being explored and telescopes being used in space and on the far side of the Moon, we shall know whether we are alone in the galaxy, and perhaps the universe. But until then, the question remains open, with a very strong bias towards the possibility that we have company.

. . . And how to communicate with them

NAPOLEON lost the Battle of Leipzig and his eastern empire in 1813 through failure to understand a coded message from Marshal Augereau that his troops were too exhausted to build a bridge. The lesson is plain: coded messages are only useful if their recipients can decode them.

This is a problem which has long fascinated the searchers for other intelligences in the universe. When we have determined on a star, or preferably a group of stars where there is a resonable chance of finding an intelligent planetary civilisation, how should we send the message: "We are here. We are intelligent"?

In short, what should we say to these supposed beings, who will be unfamiliar with any known language, that will not only give information about ourselves but whose meaning will also be immediately plain?

Several famous interstellar messages have either been postulated or sent. Perhaps best known is the plaque launched to the stars on the unmanned craft Pioneer 10 in 1972 which, drawn by Prof. Carl Sagan and his wife Linda, and Prof. Frank Drake, depicted a naked couple and showed the location of the earth in space.

But the Pioneer spaceships are leaving the solar system at only 26,000 mph, and their chances of being picked up by an alien in the first 10 million years of their journeys are fairly remote.

A message will have the best chance of success if sent by radio at the speed of light, which is nearly 26,000 *times faster* than the Pioneer and Voyager spacecraft.

In 1974, Frank Drake actually sent such a message, from the American radio telescope in Puerto Rico. But for us, the only sadness is that he aimed his radio beam at a large cluster of stars some 24,000 light-years distant, so that we cannot possibly get a reply from that cluster for at least 48,000 years.

Since radio is the easiest and most economical medium, it would be best to follow Drake's custom and use the binary code which enables the recipient to unravel the message in the form of a picture.

For instance, the message 1001111, 1001000, 1001000, 111111, 0001001, 0001001, 1111001 means, in the binary code: "BEWARE! WE ARE NAZIS!"

Why? It is simple. There are 49 noughts and ones in the above message. We therefore construct a grid of 49 squares, i.e. 7 by 7. The first digit on the first line is 1, so black it in. The second is 0, so leave it blank. When all the ones are blacked in, we have a drawing of a left-handed swastika.

Mr. A. T. Lawton, writing in *Spaceflight* in July, 1971, proposed a drawing of a suited astronaut made by the black and white squares of a rectangular grid with 59,200 squares.

All this speculation has been inspired by a remarkable discovery to the effect that the number of habitable planets in the universe and hence the number of civilisations, may be higher than we thought.

Dr Robert S. Harrington, of the U.S. Naval Observatory, made a computer analysis in which he pretended that the planet Jupiter was a star of the sun's mass. To his surprise, the earth's present orbit was more or less unaffected by the stronger gravitation.

In other words, the galaxy's innumerable two-star systems may after all be suitably placed regions of planetary life. Let's start working on those interstellar messages now.

Using gravity as a lens

WHILE staying in the vicinity of the solar system, how difficult would an astronomer find it to construct a detailed map of an earth-sized planet in orbit round the star Regulus, 84 light-years away, showing its continents and oceans?

A scientist's first reaction would be that the feat was impossible, no matter how powerful a telescope was used. The stars, let alone any earth-sized planets in their orbits, are so far away that they will always appear to us as dimensionless dots (although the brightest ones can be distorted by the passage of their light through our atmosphere). Who therefore is hopeful enough to assert that it will soon be possible to map not only the stellar surfaces but also their smallest planets?

The optimist is Dr Von R. Eshleman, of the Centre for Radar Astronomy at Stanford University, California, who proposed in the journal *Science* on Sept. 14 (1979) that we could magnify the light images of distant celestial objects tens of millions of times by means of a *gravitational lens*.

A "gravitational lens" is not something that can be made in a factory. The concept involves using the gravitational field of the sun's mass actually to magnify the light-signals from distant stars.

It is fairly well known from Einstein's 1916 general theory of relativity that a beam of starlight passing near the sun is bent by the sun's mass. During a total solar eclipse, when stars can be seen in the daylight sky, stars close to the sun's edge appear to change their positions. We can see stars that ought to be hidden from us by the eclipsed sun. The sun's gravitational field has deflected their light-beams.

It is less well known that it also makes them brighter. A large mass like the sun's magnifies as well as bends, and, to a certain point, the further one is from the sun, the greater the magnification of some of the stars directly behind it.

There exists a focal point far out in space from which objects seen behind the sun would be magnified about 20 million times. Astronomers who wanted a close look at that hypothetical planet in the orbit of Regulus would send out unmanned observatories to this focal point, and when Regulus was directly behind the sun they would obtain the terrific magnification.

Obviously, the further one travels from the sun, the smaller it appears. To an astronaut standing on Pluto, our most distant planet, the sun would appear merely as a bright star, its disc indistinguishable except through binoculars.

But even the orbit of Pluto, 3,670 million miles from the sun, is not far enough to make use of a gravitational lens. Even at this distance the sun's apparent size would prevent any useful magnification of Regulus.

The flying observatories must keep going until they reach a point *where the apparent diameter of the sun becomes less than the degree of deflection of the image of the star behind it.*

If this sounds complicated, try a simple experiment. Hold your clenched fist in front of your eyes, blocking your view of an object at the far end of the room. Then move your fist towards the object. Your fist appears to shrink, and when your arm is fully extended you can see the object all around it.

Your fist represents the sun, and the object Regulus. As the former shrinks, the latter appears ever larger, the magnification of its image in this case swelled by the sun's mass.

It is Dr Eshleman's achievement to bring this strange-seeming idea into the realms of practical science. He calculates that the ideal focal point would be a point 200,000 million miles out into space, 50 times further away than Pluto, a distance which light, moving at 186,000 miles per second, takes two weeks to traverse.

With somewhat more advanced versions of the technology of the Pioneer and Voyager spacecraft, which have undertaken long-term journeys to the outer planets, a spacecraft cruising at 1·3 million mph could reach the sun's focal point within about 17 years. It could then study its target-object in infra-red light, to avoid distortion from the sun's optical light.

An interesting mission would be to look at the great galaxy in Andromeda, a group of stars half again as numerous as our own Milky Way. It should be possible to construct a map of the stars in the

Andromeda galaxy which would be as much as two miles wide—a fantastic prospect.

Gravitational lenses may provide the astronomy of the next century. Magnifications of up to 20 million times would bring about a mighty leap in our knowledge of the universe.

Little green men in the Lords

THE EARL OF KIMBERLEY: My Lords, does the noble Lord not think it conceivable that Jodrell Bank says there are no UFOs because that is what it has been told to say?

LORD HEWLETT: I certainly think it inconceivable—absolutely and completely inconceivable . . . I am sorry, the existence of UFOs is even more fanciful than Gilbert and Sullivan's Iolanthe—charming indeed, but I am afraid a joke upon your Lordships' House.

Hansard Jan. 18, 1979

IN SUCH elegant old-world language did the House of Lords debate recently the possible existence of Unidentified Flying Objects. These mysterious machines were declared by some peers to be spaceships from another solar system crewed by alien explorers.

The large number of people in the public gallery, the chamber more than two-thirds filled, and the extraordinary furore when allegedly strange objects appeared over New Zealand—all testify to the obsessive interest which many people have in "Ufology."

Since the members of the House of Lords are but a microcosm of the public, I will try to convey the flavour of this debate by selecting a few short passages of it from Hansard. The ignorance of science shown

by some peers, and the good sense of others, combine
to form a curious mixture.

The Earl of Clancarty: An informed public is a
prepared one. It is on record that landing reports are
increasing all the time. Just suppose that UFOs
decided to make mass landings tomorrow in this
country—there could well be panic here, because our
people have not been prepared.

Lord Trefgarne: I have flown some 2,500 hours as a
pilot, but I have never seen a UFO. Since time
immemorial, man has ascribed those phenomena that
he could not explain to some supernatural or extraterrestrial agents. But as science has advanced, these
phenomena are understood more fully. Today, no one
takes witches seriously, and there are no fairies at the
bottom of my garden.

The Earl of Kimberley: The human mind cannot
begin to comprehend UFO characteristics: their propulsion, their sudden appearance, their disappearance,
their great speeds, their silence, their manoeuvres,
their apparent anti-gravity, their changing shapes. . . .

Viscount Oxford: We should have a worldwide
organisation to look into this matter—and why should
not we be the leaders of it?

Lord Davies of Leek: Ordinary little people have
sometimes been laughed at, especially those concerned in the famous sighting at Pascagoula in Missis-
sippi, when one little fellow fainted when he saw a
chap with one leg jumping towards him with a wizened
and wrinkled face, with pointed ears, crab claws for
hands, slits for eyes and holes beneath the nostrils
—are we right to call these men liars, hallucinators or
sensationalists? . . .

The Lord Bishop of Norwich: All the far corners of
the creative world, right out further than we can ever
see or even know by radio, are within the plan of the
Creator. I believe they are within the majestic purposes of God.

Lord Gladwin: UFOs led to the only known joke

perpetrated by Mr Gromyko (the Soviet Foreign Minister) who said: "Some people say these objects are due to the excessive consumption in the United States of Scotch whisky. But it is not so. They are due to a Soviet athlete, a discus-thrower practising for the Olympic Games and quite unconscious of his own strength."

Lord Rankeillour: These machines are potentially dangerous. They give off blinding light and crippling rays. They start forest fires, eradicate crops and cause great distress to animals. If the British population was aware of this, they could take precautions.

Lord Hewlett: Two days ago, I was briefed by Sir Bernard Lovell and his senior staff at Jodrell Bank (radio telescope). Of all the thousands of reports of sightings that have been made, whenever it has been possible to make an investigation, they have been found to be natural phenomena, or in some cases, I regret to say, pure myth. Jodrell Bank has been on watch for 30 years, it has found nothing whatsoever to report about UFOs. For the noble Lord, Lord Davies of Leek, to throw at me "some scientists think . . ." is not good enough.

The Earl of Cork and Orrery: With respect to my noble friend, the fact that the Jodrell Bank telescope has not seen something not only does not prove, but is not even particularly good evidence, that it was not there . . .

Lord Strabolgi (for the Government): Where are these alien spacecraft supposed to be hiding? Ufologists have had to claim that the aliens are based in the depths of the sea, or even in a great hole in the earth. But there is nothing to convince the Government that there has ever been a visit by a single alien spacecraft, let alone the numbers of visits which the noble earl, Lord Clancarty, claims are increasing all the time. . . .

As Private Willis might have said: "When all night long a chap remains on sentry-go, he will see practically anything."

The unkindest cuts of all

ONE OF the most important lobbies of modern times is now being mounted in Washington. A Congressional sub-committee has chopped the equivalent of £700,000 from a £1 million publicly-funded project to search for signs of intelligent life elsewhere in the universe, and scientists are urgently trying to get the decision reversed by the Senate. The remaining £300,000 in the words of a NASA official, "will not even be enough to start design work on a radio antenna."

The £1 million, were it forthcoming, would be spent on the construction of more sensitive radio receivers and ingenious new computerised data-processing systems to search among the approximately 180,000 million stars of our Milky Way galaxy for artificial signals or any sign of technology more advanced than our own.

This is an incredibly modest investment for a project which, if successful, could change man's whole outlook. The proposed expenditure amounts to less than 0·0001 per cent of America's gross national product, and is a mere fifth of the sum which Britain is now spending on a probably fruitless programme of research into the viability of certain bizarre energy sources.

Only in the narrowest technical sense is this an internal American dispute. The issue is of such importance to humanity that foreigners are justified in commenting on the behaviour of officials concerned in it.

The opposition in the two houses of Congress to an organised search for intelligent alien life appears to centre around one man, Senator William Proxmire, who objects to it as a waste of public money.

Now it would be reassuring if I were able to report that Senator Proxmire had considered the question profoundly from a philosophical point of view, and had foreseen some danger to the human race in the event of the search being successful.

But this does not seem to be the case. Prof. Isaac Asimov expressed the opinion last week that Senator Proxmire was "bone from ear to ear." Whether this judgment is fair may be deduced from the recent statement from the senator that the proposed search was "crazy science fiction which should be postponed for a few million light-years."

Not knowing that light-years are units of distance, not time, perhaps gives a clue to this senator's knowledge of astronomical matters.

A letter in the *Wall Street Journal* made the comment: "Inanities broadcast by Sen. Proxmire are now diffusing into space at the speed of light. In a few years time they will reach the neighbouring stars, and if the intelligent beings there are beaming messages at us they will decide that their effort is not worthwhile and turn their antennas elsewhere."

"Mr Proxmire deserves an honorary membership in the Flat Earth Society," declared the distinguished astronomer, Prof. Frank Drake. A few days later, this scientist was astonished by a telephone call from Mr Proxmire's office. "The senator would like to know what the Flat Earth Society is. Is he really likely to be honoured by membership of it?"

One British organisation has done much to confuse the issue. The U.K. Cutty Sark Scotch Whisky Company has offered a prize of £1 million to anyone who can discover an alien spacecraft or artefact which has originated from outside the solar system.

A fairly safe risk for the company, you might say, although it has insured itself at a premium of thousands of pounds against having to pay out the prize. But there is a nasty catch. According to a director of the company, Mr Russ Taylor, the firm will pay nothing for the detection of an artificial radio signal from space.

Mr Taylor gave his reason at a recent meeting in London at which one scientist walked out in disgust. "Discovery of alien beings by radio signals is much less likely than by a discovery of an alien aircraft," he pronounced. "Besides, no one could prove that a radio signal from space was artificial."

This is the most specious nonsense. A genuine extraterrestrial signal would be unmistakable, since it would be intended to be understood. In short, the prize has been offered in such a way that it will be very difficult for anyone to claim it.

The conditions for claiming the prize are hilarious. Competitors must take their alleged alien spacecraft at their own expense to the Science Museum in London for verification by experts—who, incidentally, angrily deny having made any such commitment.

Now the alien mother-ship in the recent science fiction film "Close Encounters of the Third Kind" appeared to weigh about 30,000 tons and was at least 200 yards in diameter, which probably would be necessary for an interstellar voyage.

Mr Taylor could offer no advice when I asked him how such an object should be carried into central London without knocking down buildings, or whether the alien crew might object to being moved around in this way, or whether the cost of transporting the craft might exceed the £1 million prize money.

These are two classic instances of how a great scientific enterprise can be imperilled.

Where the world came from

NEWS OF the recent break-through in controlled nuclear fusion at Princeton has puzzled many people

who believed they understood the mechanics of the
creation of the world and the origin of the materials
which comprise it. The Princeton experimenters succeeded in heating hydrogen for a tenth of a second to
110 million degrees F., nearly four times hotter than
the core of the Sun, in their attempts to imitate the
Sun's power and make hydrogen "fuse" in helium.

This, indeed, is all that happens in the core of the
Sun. Hydrogen is converted into helium, releasing
energy as it does so, at a conversion rate of some four
million tons of hydrogen per second.

Nothing else is being created in the Sun. Nothing
else has ever been created there; nor will it be for
billions of years to come.

Yet this fact appears to contradict the evidence of
our senses. Hydrogen and helium are the two lightest
elements in the universe. Yet everywhere we look on
Earth, we see heavier ones.

Where do these elements all come from if they were
not made in the core of the Sun?

It used to be supposed that they were made somehow by chemical evolution during the formation of the
Earth, in the same way, perhaps, as carbon dioxide
breaks down into oxygen and carbon. But this is only
possible with compounds, not with the elements. For
one element to be transmuted into another requires a
nuclear transformation at temperatures of hundreds of
millions of degrees.

The answer to the mystery is that the elements on
Earth heavier than helium came originally from the
Sun, but they were not *made* in the Sun. The Sun isn't
hot enough. The core where helium is made is 30
million degrees F., but 200 million degrees are needed
for helium in turn to fuse into carbon, oxygen and
neon.

For neon to fuse into magnesium, 1,440 million
degrees are necessary. The production of aluminium,
silicon, sulphur and phosphorus requires a temperature of 2,700 million degrees, and 4,500 million are

needed to make still heavier elements such as titanium, chromium, manganese, iron, cobalt, nickel, copper and zinc.

Where are these fabulous temperatures to be found? Only in another star far more massive than the Sun. Such a star must have ended its life in a violent supernova explosion which itself brought about the Sun's creation. This is the modern scientific explanation of the creation of the world, and the only one which fits all the facts.

An exploding supernova, perhaps about 60 light-years away, sent blasting shock-waves of heavy elements for about 100 light-years in every direction. The fast-moving shock-waves enveloped clouds of interstellar hydrogen, one of which was destined to form part of our Sun.

The shock-waves compressed the hydrogen, adding its own materials to it. The primitive Sun then collapsed under its own increased gravitation until it became sufficiently hot in the core for thermonuclear ignition to take place. The outward-pushing force of the nuclear reaction henceforth balanced the inward-pushing gravitational force and the Sun became a stable, shining object that would last for billions of years.

During this violent convulsion, heavy elements were flung out of the fast-rotating primitive Sun, and these in turn coalesced into planets.

But the Milky Way galaxy contains about 200,000 million suns. Are we to suppose that *each* of these had its own parent supernova?

Certainly not. That would be an extremely inefficient way for nature to construct a galaxy. We must assume instead that the average rate of star construction during galactic history has been perhaps one million stars per supernova. Since supernova fragments blast out in every direction, they may be expected to compress innumerable gas-clouds in their path.

The proof of this hypothesis may be seen through a telescope. Supernovae are still making stars. Astronomical photographs show the process happening in the constellation of Canis Major.

The world's end

"I CANNOT convey the sense of abominable desolation that hung over the world . . . The sky was no longer blue. North-eastwards it was inky black, and out of the darkness shone brightly the pale white stars. Overhead it was a deep Indian red and starless, and south-eastward it grew brighter to a growing scarlet where lay the huge hull of the sun, red and motionless."
   —H. G. Wells, "The Time Machine" (1895)

Wells, like the turn of the century astronomers whose ideas inspired this part of his great story, knew little about the workings of the sun, and of how the world will ultimately end.

As his Time Traveller moved on his machine millions of years into the future, he encountered a colder and darker earth and a fading sun that was running out of its known fuel. The red, dying sun emitted little heat. It was snowing steadily and there were fringes of ice along the sea shore.

Very different would the Time Traveller's adventures have been if Wells could have written his book with today's knowledge. For the sun will not fade out like an exhausted log fire; it will expand until it has increased its volume 40 *million times*, and the earth will be swallowed up inside it.

These fearful events are unlikely to occur for another 5,000 million years, but the world's fiery doom is unavoidable.

Time after countless time, the glaciers will advance

inexorably from the poles, gouging out great valleys. The oceans will continuously change their sizes and their shapes.

But these changes will be endurable. The first but least serious threat will come from the moon. The earth's rotation is even now being slowed by the moon's tidal drag, and each day is on average 25 thousand millionths of a second longer than the last.

There will come a time when each day and night will last for 18 hours apiece. Days will be hotter and nights colder, with catastrophic effects on crops and climate. The moon, which is now receding from the earth at about one foot every 30 years, will acquire the rotational energy which the earth is losing, and it will shine ever more faintly in the sky.

But the tidal fields of the sun will again speed up the earth's rotation, overcoming the lunar drag, and the moon will be pulled inwards once more. This time it will approach dangerously close to the earth, perhaps to 9,600 miles—2·44 times the earth's radius—the theoretical limit at which the moon must break into fragments under tidal pull, and form a ring system round the earth.

What effect this orbiting ring of giant stones may have on life on earth we do not know.

But the sun meanwhile will be preparing itself to inflict disaster. The great nuclear furnace in the core, which radiates energy by converting hydrogen to ·helium, is piling up its own nuclear wastes in the form of inert helium ash.

Ever more hydrogen will be needed to keep the sun shining at its present strength. At length all the hydrogen in the core will be used up. Hydrogen from the outer layers will pour into the core to keep the reaction going.

The helium ash will get steadily hotter. At length it will reach the critical temperature of 200 million degrees F, and the fatal Helium Flash will occur. As the helium itself ignites, the explosion will tear apart the outer layers within a space of hours, and the blazing central core of the sun will be revealed.

The outpouring of X-rays and gamma rays that follows this catastrophe will destroy all life remaining on the world's surface, although underground habitation will still be possible.

Even this desperate expedient will not be safe for long. In the 30 million years that follow the Helium Flash, helium atoms will continue to merge explosively to form oxygen, carbon and neon. The sun will expand into what astronomers call a "red giant." Its diameter will expand from its present 865,000 miles to some 300 million, swallowing and destroying the inner planets.

Another 50 million years will suffice to complete this nuclear orgy. Helium supplies will run out. The bloated sun, unlike more massive stars, will not be hot enough to burn the carbon and other elements, and there the drama will end.

The sun will collapse until it is far smaller than it is today. It will remain a superdense object made largely of crystalline carbon, or pure diamond, no larger than our world which it will have devoured, gradually cooling until the end of time.

Imagine a Time Traveller observing this scene from a safe vantage point, perhaps from the newly discovered moon that orbits distant Pluto.

The sky will no longer be filled with stars, since star formation will have ended in this region of space, and all the familiar bright stars that make up our constellations will long ago have disappeared. New stars and galaxies may be shining far away—but too far to be seen with the naked eye. Empty space will appear illimitable and black night will reign supreme.

Star-gazing with binoculars

In frosty winter and in balmy summer, the amateur astronomer can see many interesting things in the sky, armed only with a tripod and a pair of binoculars.

People deciding to take up astronomy often make the mistake of spending £50 or more on a small telescope, on the grounds that because it magnifies up to 75 or 100 times, it must reveal far more information than binoculars. However, their field of vision is so narrow, you can never be sure that you are seeing the star you are trying to look at, and unless you have mechanical controls to compensate for the earth's rotation, the star will rush through the eye-piece and you will quickly lose sight of it.

Binoculars have none of these problems. They magnify only between about 7 and 20 times, and a star therefore scarcely moves while you are looking at it.

But they scan a far wider area of the sky than a small telescope. It is therefore easy to glance up at a particular object with the naked eye, and then train the binoculars on it. This gives one an awesome feeling that one is out in space, personally communing with the universe, that a small telescope can never give.

There are four main objects which it is easy and worthwhile to find in the night sky. The binoculars —the more powerful they are the better—must be rigidly fastened to the tripod, and the tripod should be sturdy and should have knobs enabling smooth movements to be made up and down and from side to side.

The stars, on a moonless night, at first sight resemble chaos, and nothing is clearly distinguishable no matter how much astronomical theory one has read. An excellent accessory therefore is a plastic disc-like device called a Philips' Planisphere, which shows the positions of all the northern constellations at different dates of the year and at different hours of the night.

This should be supplemented with a star chart, which maps the heavens in considerably more detail. *The Daily Telegraph* Sky at Night Map is very good, and *Norton's Star Atlas* (Gall and Inglis, Edinburgh) is even more elaborate.

Now the first stars to appear on a summer or early autumn evening will be the brilliant triangle of Deneb, Vega and Altair which straddles the Milky Way.

When they have been picked out, the bewildering chaos of the sky becomes clear and with the use of the Planisphere and the Atlas, all sorts of other objects can be sought and found.

My first choice for spectacular beauty is the Pleiades cluster in Taurus. Known as the Seven Sisters (one of them has faded a bit) it consists now of six brilliant stars very close together when seen with the naked eye, and several hundred when studied through binoculars.

Perhaps nothing in the sky is easier and more satisfying to see than the Pleiades. James Pickering writes of them in his 1001 *Questions Answered About Astronomy* (Lutterworth Press): "The whole cluster blazes with stars in glorious profusion, in streams and in loops and in all possible combinations."

The next feat is to relive the triumph of Galileo without having to endure the terrors of the Inquisition. Find the giant planet Jupiter and see for yourself its four largest moons in orbit around it. That three of them are larger than our own moon will be far from obvious, since they will appear as tiny points of light. But Galileo found them with one of the world's earliest and crudest telescopes, and so it should be easy with modern binoculars.

Jupiter is very easy to spot once it has risen, since it is the fourth brightest object in the night sky, beyond the Moon, Venus and Mars. Through binoculars of 15 magnifications or more it becomes a clear, orange-coloured disc.

The Moon is also a fine sight, with its magnificent craters and its two main mountain ranges, one of which rivals the Himalayas in height.

Perhaps the greatest achievement of all is to see the great spiral galaxy in Andromeda. A small fuzzy blob of light is all you will glimpse of it through binoculars, but its true nature is mind-blowing. It is one of our neighbouring galaxies and it contains *no less than* 250,000 *million suns.*

The Andromeda galaxy is tricky to find, because of its faintness. One quite easy way is to follow a leftward line from Vega through Deneb. When the length of the line to the left of Deneb is about one-and-a-half times the Vega-Deneb distance, the galaxy will be somewhere near the end of it. It can be a tedious search but the result is worth it.

This has been strictly a guide for the low-budget amateur. A really good telescope, such as a six-inch reflector with mechanical controls that costs several hundred pounds or a £400 Questor with camera attachments, is far superior to any binoculars.

Sunspots and the climate

MANY PEOPLE must be wondering whether there is any connection between the violent explosions that have been taking place on the surface of the sun and the weather which, for many of us, made the summer of 1980 one of the most miserable for years.

It might seem a reasonable question. The incidence of violent solar flares is now approaching the climax of its 11-year cycle. Streams of hydrogen hundreds of thousands of miles long have been shooting out from the sun, disrupting short-wave radio broadcasts on earth and making electronic instruments go haywire.

But any link between the sun's activities and our day-to-day weather is so far undefined. The world's weather systems vary so much and are so complex in relation to each other that to predict accurately tomorrow's weather on the basis of today's sunspots is at present impossible.

When we come to climate, however, the picture is different. Long periods of balmy weather appear to coincide with those of violent solar activity as in the present century, and very cold periods, or "little ice ages," seem to happen when the sun is relatively quiet.

The discovery of this connection makes a fascinating detective story. In 1893, E. Walter Maunder, an astronomer at the Royal Greenwich Observatory, was browsing through old journals and he could hardly believe what he was reading.

He found that between 1645 and 1715 there were virtually no spots on the sun's surface at all. The sun, during this 70-year period, was in other words behaving quite differently from today when spots appear continually. It still shone warmly, but it lacked that extra energy which produces the spots.

This 70-year period, now known from its almost complete lack of sunspots as the Maunder Minimum, corresponded with a very cold period in Europe and America.

Alpine glaciers advanced further than at any time since the last great Ice Age 15,000 years before. The Viking colony in Greenland was cut off by drifting pack-ice which failed to thaw, and oxen were frequently roasted on the frozen Thames in London.

In 1671, the editor of the *Philosophical Transactions of the Royal Society* made an announcement that sounded quite commonplace at the time, but which we today would think extraordinary.

"At Paris," he wrote, "the excellent Signior Cassini hath lately detected again spots on the Sun, of which none have been seen these many years that we know of.''

There is no reason to doubt Cassini's report. Giovanni Cassini was an opinionated, self-important man, but as an astronomer he was completely reliable.

Now for the evidence of the tree-rings of the bristlecone pine. The oldest of these mountain trees, which are found in western America, is about 7,000 years old.

Their rings show the changing abundance, year by year, of the radioactive isotopes carbon-14, which strike the earth in cosmic rays from our Milky Way galaxy. The evidence from the rings is explicit. When the sun is quiet, as during the Maunder Minimum, there is an abundance of carbon-14. When the sun is active in periods like today's there is much less of it.

The reason for this is obvious. A ''noisy'' sun emits a strong magnetic field which tends to blot out the cosmic rays, while the weak magnetic field of a quiet sun allows more of them to strike the earth.

From the tree-rings, therefore, we have an accurate picture of the sun's behaviour going back 7,000 years. And whenever historical evidence about the climate is available the quantities of carbon-14 appear to correspond to cold and warm periods.

The 15th century, for instance, marked the Sporer Minimum of sunspots (named after a German colleague of Maunder's) and was cold. From the middle of the Dark Ages (about 600 AD) until about 1350 was the warm period of the Mediaeval Maximum.

There was a warm Roman Maximum covering most of the period of the empire, while the Greek and Homeric epochs were a minimum and cold. A detailed chart of all these warm and cold periods is given on Page 88 of the May, 1977 *Scientific American.*

The cause of sunspots and the absence of them is unknown. Something peculiar happens at long, irregular intervals in that vast thermonuclear reactor in the core of the Sun, but what it is nobody is sure.

Yet there is one thing we can be sure of. The present era of sunspot activity, which comes and goes in mini-cycles of roughly 11 years, is rightly called the Modern Maximum. Until the next minimum begins there is no prospect of another Ice Age, and those who try to roast oxen on the Thames will probably drown.

The planets in sequence

THE ancient mathematical sequence known as the Fibonacci Numbers has begun to show such wide-spread applications in natural science, psychology, and now planetary astronomy, that many scientists are beginning to wonder whether it may be linked with some mysterious law governing events throughout the universe.

The Fibonacci system was the discovery of Leonardo Fibonacci of Pisa in 1202, and was ignored through the centuries until its extraordinary capabilities began to interest a group of Californian mathematicians in 1960. In its simplest form, the sequence goes 1, 1, 2, 3, 5, 8, 13, 21, 34, 55, 89, 144, 233, 377, and so on, so that each number is the sum of the two previous ones. Here are some of the system's many peculiarities.

The spiral seed formations of sunflowers are nearly always found in Fibonacci numbers. Ordinary-size sun-flowers have 34 or 55 spirals growing in opposite directions, while giant ones usually have 89 and 144. Tree branches grow in a Fibonacci sequence, as also do the scales on fir-cones.

The ratio between any Fibonacci number beyond 3 and its successor is always the famous irrational number 1.61803, which is obtained by halving the sum of 1 and the square root of 5. This number's relation-ship with 1 has throughout architectural history provided the Golden Rectangle, of which examples range from the Parthenon to the most attractive of modern blocks of flats. Leonardo da Vinci, among others, found that this geometric combination could give effects especially pleasing to the eye.

Every third Fibonacci number is divisible by 2, every fourth number by 3, every fifth number by 5, every sixth number by 8, and so on. The divisors themselves form a Fibonacci sequence in which each is the sum of its two predecessors.

Fibonacci first used the sequence to predict the population growth of his pet rabbits. He assumed his single pair of rabbits would produce a new pair every month. Each new pair would do the same after taking a month to mature. The rabbits conveniently reproduced in pairs only, and by the end of the year, as predicted, the total was 377, the 14th Fibonacci number.

The Fibonacci Association a small group of mathematics teachers at St Mary's College near San Francisco, publishes a *Fibonacci Quarterly* for its 900 members. Having feared at first that serious scientists would dismiss them as a "bunch of nuts," they have now become accepted by the mathematical community.

Perhaps the most remarkable instance of Fibonaccism is its recently discovered improvement on Bode's Law, which governs the distances of our planets from the Sun. Johann Bode found in 1776 that when 4 was subtracted from each distance (measured in "astronomical units," known as A Us, where Earth's distance equals 1) they would climb in a simple doubling sequence of 3, 6, 12, 24, etc. But it has now been found that the Fibonacci system works even better—albeit somewhat roughly—without the contrived complication of subtracting 4.

"Fibonaccially" speaking, the distances go: Mercury 0.4 A Us; Venus 0.7; Earth 1; Mars 1.5; the giant asteroid Ceres 2.8; Jupiter 5.2; Saturn 9.5; Uranus

19.2; and Neptune 30.1. Infuriatingly, Pluto breaks the sequence entirely with only 39.5. But it has been argued that Pluto is not a proper planet anyway, just an escaped moon. The great advance is the discovery that Neptune conforms with Fibonacci, whereas it disobeys Bode's Law. To explain this discrepancy, a few of Bode's followers have tried to pretend that Neptune also is a non-planet, a thesis which has convinced nobody.

The new sequence may explain why we have found no more Solar planets lately. They will be located far out in interstellar space. Our 50th planet, if it exists, will be at a distance from the Sun of 93 million miles (1 A U) multiplied by the 50th Fibonacci number, which gives a number in miles of 1 followed by 18 zeros.

RELATIVITY AND THE UNIVERSE

Let us move now into the bizarre world of relativity, where twins can be of different ages, and where space and time can cease to exist. Even stranger are the predictions of quantum mechanics, which hint at parallel worlds identical to our own and the possibilities of altering the past.

Youthful journey into space

THE prediction that an astronaut who travelled into space at a velocity close to the speed of light could return to Earth and find that he was literally younger than his own children provides one of the most beautiful paradoxes of modern science.

I had hoped that it was also one of the most widely known. But sadly this is not so. Only two weeks ago, a quite literate gentleman called Mr T. Crosbie wrote to the *Economist* dismissing this paradox in a few lines, and added: "I don't believe it. Does anyone?"

I will be doing a public service if I answer Mr Crosbie's rhetorical question.

Albert Einstein predicted in 1905 that the faster a man travelled in relation to the Earth the more slowly he would age. As he approached nearer and nearer to the speed of light, which is 670 million m.p.h., time on board his spaceship would increasingly slow down in comparison with that of his twin brother back on Earth.

Although it has eluded Mr Crosbie, the proof of this statement is almost childishly simple:

It was discovered in the late 19th century that light-beams in space behave in the most peculiar way. They *always* move at the same speed—670 million m.p.h., irrespective of the speed of their source.

Consider this in the case of an astronaut whose ship is rushing towards Planet X at half the speed of light. He might imagine, if he had never heard of Einstein, that the light from the sun of Planet X was hitting his spaceship at $1\frac{1}{2}$ times the speed of light.

And so he gets out his measuring instruments to see whether this is true. To his astonishment he finds that it is not. The light from that star is hitting his ship at precisely the speed of light and no more.

If he has a logical mind, there is only one possible conclusion that he can draw from this, namely that his clocks are running slowly.

Now Mr Crosbie, who is clearly no fool, gives the following provocative answer to the paradox.

Suppose, he says, that the astronaut has on board a telescope of fabulous power which enables him to watch the receding face of Big Ben as he speeds away from Earth at near the speed of light.

Now the further he gets from Earth, the longer the light signal from Big Ben will take to reach him. And so the faster he goes the slower will time appear to him to be running.

Fine, says Mr Crosbie. But what happens when he turns around and heads for home. Will he not see the hands of Big Ben swinging round at a furious pace, making up for their previous slowness, and bringing him back to Earth exactly the same age as his twin brother?

But Mr Crosbie has forgotten Planet X, the place which the astronaut has been visiting! Suppose that this planet also has another clock which is equally visible through the astronaut's amazing telescope.

As he looks backwards at this clock during his return journey, he sees the same thing happen as when he watched Big Ben on the way out. Looking backwards at Planet X, his time seems, quite correctly, to be slowing just as it slowed on the way out.

Now because, in this example, the pace of time in London and Planet X are always the same, there must be something wrong with Mr Crosbie's answer.

It is obvious where he has gone wrong. The signals from Big Ben cannot reach him any faster than the speed of light, no matter how fast he is rushing towards Earth. And so while he is travelling fast, he will always be ageing much more slowly.

One might of course say in objection that the Earth was moving while the astronaut stood still. If this were so, why should the astronaut age more slowly than his friends on Earth? Einstein's devastating answer is that when the Earth moves, the entire universe moves with it, but when the astronaut moves, he moves alone.

The Earth requires no engine to keep it in orbit round the sun. But the astronaut has used an engine to accelerate; and it is his acceleration which causes his time to slow down.

The slowing of time in an accelerated vehicle has been proved by experiment. In 1971, two scientists, Joseph Hafele and Richard Keating, flew round the world in a jet aircraft carrying with them a super-accurate atomic clock synchronised to an identical clock on the ground.

On their return, the airborne clock was found to be running a tiny fraction of a second behind the ground clock. The aircraft speed of 600 m.p.h. is little compared with the speed of light, but the finding was enough to prove Einstein was right.

Einstein's way with the void

LET US look at what I will call the Paradox of the Dictator and the Red Carpet.

It has nothing to do with politics; it is rather a device I have invented to further understanding of that most awesome discovery of modern science, Einstein's general theory of relativity of 1916.

It has been said of the theory that never were predictions made that were easier to understand and at the same time more difficult to believe. The human mind naturally finds it hard to grasp the concept of absolute "nothingness." Let me try to explain.

A reader of this newspaper, Mr Paolo Contarini, wrote asking the main questions which the general theory answers: "What is the end of the universe, and what is beyond the end, and so on *ad infinitum*?''

He asked again in a second letter: "How did matter itself originate? How did space come to exist at all? These are the great fundamental mysteries which puzzle ordinary people." Let us try to answer the question with an anology.

Picture a furious-looking Mussolini disembarking at a railway station, a scene in Charlie Chaplin's film "The Great Dictator."

Mussolini is carried to Berlin by an incompetent engine-driver who keeps on stopping at the wrong place on the platform so that the two men carrying the red carpet have a difficult time unrolling it at the exact place where he will disembark.

So hasty have been their preparations, they are still unrolling the carpet as Mussolini walks on to it. Conversely, if one ran this scene from the film *backwards,* they'd be rolling it up as he vanished into the train.

The dictator, in this analogy, represents physical matter, that is to say the stars and galaxies, while the carpet is time and space. Just as in the theatrical world of fascist diplomacy, a dictator could not appear without a red carpet, nor vice versa, so, in Einstein's revelation of the universe, space and time cannot exist alone; they can only exist in the presence of matter.

It is well known that the universe is expanding. That is to say, the galaxies are rushing apart from each other. The question therefore arises: *into what* are the galaxies rushing?

Before 1916, it would have been said that they were occupying hitherto empty space. But Einstein showed that there could be no such thing as empty space. Space and time by themselves cannot exist. Matter creates the space and time into which it advances, just as the presence of Mussolini (in a manner of speaking) creates the red carpet on the platform.

To put this another way, a spaceship could not travel to a place beyond the universe because *there is no such place.*

Reading this apparently meaningless statement, many people of strong common sense will throw up their hands in disgust and say that Einstein's equations are for mystics and poets. How can there be a place where there is no place?

To put it as simply a possible, space and time must be regarded as physical entities which are *curved* by the gravitational fields of massive objects, such as stars and galaxies. Take away these massive objects and nothing whatever would remain. Asking what is beyond the universe is therefore like asking what is north of the North Pole.

The same applies to the Big Bang which began the universe some 18,000 million years ago. It is believed that all the primeval matter which now comprises the universe burst out of another dimension in one cataclysmic explosion, and it has continued to rush outwards ever since, creating its own time and space as it does so.

And so there is no question of anything having happened "before" the Big Bang, since it was the beginning of time. Nor could we say that it happened in any particular "place." Since it was also the beginning of space, the Big Bang happened *everywhere.*

Many people will say that the general theory cannot be more than the wildest speculation, in a slightly poetic vein. But this is not the case. It was proved by experiment in 1919, and many times since.

Again and again it has been shown, usually during the convenience of a solar eclipse, that light from a distant star is bent by the mass of the Sun, and that it reaches the Earth in a slightly curved path. Even radio signals from the Viking instruments now on Mars showed fractional delay caused by a curvature when Mars was near the Sun. Why? Because light-rays and radio signals must follow the paths of curved space.

The general theory describes the universe with a beauty that no rival concept can match. In the words of Prof. John A. Wheeler of Princeton, a co-inventor of the H-bomb: "No purported inconsistency with its predictions has ever stood the test of time. No logical inconsistency in its foundations has ever been detected. No acceptable alternative has ever been put forward of comparative simplicity and scope."

Einstein and Sherlock Holmes

"YOU will never amount to anything," a scornful Munich high school teacher once told a boy whom he considered lazy, absent-minded and rebellious. The prediction proved wrong, for the boy in question was Albert Einstein.

Einstein, born in Germany in 1879, changed the face of the 20th-century profoundly, and we have yet seen but the smallest part of the fruits of his discoveries. It is hard to live in the modern world without noticing some sign of them. Television-picture tubes, solar panels on spacecraft and photo-electric-eye devices which open doors to us—all owe their existence partly to his discovery in 1905 that light-beams consist of separate photon particles.

The laser beams, which have a host of uses ranging from electronics to medicine, would have been impossible without a paper which Einstein wrote in 1917, laying down the basic principle of the laser.

But it is for his bizarre and awesome theories of time and space that he is most famous. He transformed the orderly universe deduced by Isaac Newton. According to a well-known couplet:

Nature and Nature's laws lay hid in night:
God said "Let Newton be!" and all was light.
It did not last. The Devil howling, "Ho!
Let Einstein be!" restored the status quo.

But Einstein's two theories of relativity are not all

that frightening. Almost anyone can understand them—and then be astonished that he ever thought they might be difficult.

They centre around the behaviour of light. It was found at the end of the last century, to people's great surprise, that light rays in the vacuum of space behave in a very curious manner indeed. *They always travel at the same speed, 670 million mph, irrespective of the speed of their source.*

This means, quite simply, that if we were in a spaceship rushing towards a star at half the speed of light, the light from that star would not reach the ship at $1\frac{1}{2}$ times the speed of light, as commonsense would lead us to expect; it would arrive at exactly the speed of light!

From this fact, Einstein made three deductions:

That time inside a spaceship moving at close to the speed of light would actually slow down, so that its occupants would age more slowly.

That the ship's length would contract in the direction of motion.

That the energy needed to accelerate the ship would rise towards infinity.

These remarkable conclusions would have been obvious at least to Sherlock Holmes, that most avid of *Telegraph* readers, and one can imagine the conversation which would follow.

Holmes put down his paper. "An interesting line of reasoning, eh, Watson?"

"My dear Holmes, it is quite absurd. How could this man Einstein have made these extraordinary deductions about time slowing down, and all the rest, just from the fact that the speed of light is constant? It's utter twaddle!"

"You are too timid, Watson. The facts are before you, but you do not use them. Suppose we have an experimenter on board the ship. An unimaginative man even. That blockhead Lestrade, perhaps. He measures the speed at which the star's light reaches the ship, and finds that it is exactly the normal speed of light. So what does he conclude?"

"I cannot imagine."

"Why, that his clocks must be running slowly! Surely it is elementary? And the shortening of the ship's length and its increase in weight follow by the same reasoning."

"Well, now that you explain it . . ."

"You see how absurdly simple it is. That is Einstein's special theory of relativity. Now, Watson, did you by any chance hear a loud crash in this room last night?"

"Yes! I thought Mrs Hudson would have hysterics."

"I had been musing about Einstein's general theory of relativity. I was doing an experiment with the coal scuttle. I tied a rope to it and whirled it round my head. Unfortunately the rope broke, and the coat scuttle crashed into the dresser. Now what sort of force would have propelled the object with such violence?"

"Why, it was centrifugal force. You were whirling it round your head."

"It is always a mistake, Watson, to use two words where one will do. Why not call it gravity?"

"But gravity and centrifugal force are not the same thing!"

"And what, pray, is the difference?"

"Well, I suppose they *appear* to be the same thing, but . . ."

"Exactly. They appear to be the same thing because they *are* the same thing. The coal scuttle was moving relative to the mass of the universe. Therefore it felt the pull of gravity. Gravity is part of the fabric of space-time, brought into existence solely by the mass of the universe."

"Holmes, this is preposterous! Is Einstein really saying that without the mass of the universe there would be no gravity?"

"And no time or space either! There would be nothing, absolutely nothing. Mass, gravitation, time and space all depend on each other for their existence. I had thought even you might have deduced that."

Dr Watson was at the window. "Holmes, we have visitors."

Holmes peered out. "Ah, yes. The Prime Minister and the Home Secretary. I fancy Moriarty has stolen a nuclear weapon. A pity, perhaps, that some of Einstein's researches were not kept secret."

"You mean his discovery that the energy contained in a substance equals its mass multiplied by the square of the speed of light? That $E=mc^2$?"

"Precisely. Now we must prevent Moriarty from making too practical an application of Einstein's theories."

It's earlier than you think

CAN TIME run backwards? The question at first sight might sound too preposterous to discuss. But physicists do not take this view. They are becoming increasingly convinced that in certain conditions it must do so.

It is well known already that clocks slow down in a vehicle moving at an extremely high speed. At the speed of light itself, which is 670 million m.p.h., time stops altogether; a person who was miraculously able to traverse the Universe riding on a light beam would make the entire journey in one eternal moment.

At speeds faster than light, according to Einstein's equations, time does not merely stop; it actually runs backwards. This has led the American physicist Gerald Feinberg to propose the existence of the "tachyon" particle, which *always* travels faster than light, and never slows down and therefore moves always from the future to the past.

The matter has been taken still further by Dr P. C. W. Davies, a distinguished mathematician at King's College, London, who suggests that time does run backwards in the next cycle of the Universe, that is to say some 80,000 million years from now, when all the matter in the cosmos has collapsed on itself and re-expanded.

Indeed, he has suggested, it would be unnatural if it did not. For time to flow only in one direction would be "unsymmetrical." It would violate Isaac Newton's law that "every action has an equal and opposite reaction."

"Our everyday concept of past events 'disappearing' and future ones 'coming into being' has no place in physics." Dr Davies explains.

It would be as absurd, in his opinion, to take such a view of a geographical space. Nobody would suggest that north was continually disappearing and that south was always "coming into being." Just as all points on a map must always exist, so must all points in time.

Whether it represents truth or fantasy, Dr Davies's view has an honourable tradition in literature. In Plato's "Statesmen" a visitor tells Socrates of the weird events in his hometown where time runs backwards:

"Mortal nature is reversed and all men grow young. The white locks of the aged darken, and the cheeks of the bearded man become smooth. The bodies of the youths grow smaller, becoming assimilated with a newly-born child in mind as well as body. Finally, they waste away and wholly disappear."

F. Scott Fitzgerald uses a similar theme in his amusing short story, "The Curious Case of Benjamin Button." Benjamin is born a white-haired old man at the age of 70. His life follows a certain insane logic. He enters kindergarten at 65, goes through school and marries at 50.

When he is 12, the Army appoints him brigadier-general on the grounds of previous military experience when an adult. He arrives at the Army base as a small boy and is packed off home. He finally vanishes into his mother's womb.

It would be very difficult to discover a part of the Universe where time actually did run backwards. For one thing, all the stars there would be black and invisible since light would flow into them rather than from them.

Communicating with people who lived in such a place would be impossible, since they would "forget" the communication as soon as it had happened because it would be in their future.

Yet because of the essential "symmetry" of nature, the backward-flowing people would regard their way of life as being as normal as we regard ours.

Is Gerald Ford still President?

"How would you like to live in Looking-glass House?"—Lewis Carroll.

IT IS a fact well known to most people that Jimmy Carter won the 1976 U.S. Presidential election. But a number of physicists consider this statement over-simplified. They assert that another world exists, identical in most respects to our own, in which Gerald Ford is still President of the United States.

One's first reactions to this seemingly fantastic proposition is either a desire to migrate to this perhaps more fortunate world, or else to conclude that the physicists have gone mad and have abandoned physics for metaphysics.

They have not gone mad nor, in currently known science, does there seem to be any possibility of communicating with this "other world," although, as will be seen in a moment, certain laboratory experiments appear to suggest its existence.

The belief that it may exist follows from the theory of quantum mechanics, which explores the strange behaviour of sub-atomic particles, and from Werner Heisenberg's "uncertainty principle," which predicts that complete information about sub-atomic behaviour is for ever unattainable.

These ideas are lucidly explained in a recent book, "Other Worlds: Space, Superspace and the Quantum Universe" (J. M. Dent, £7·50), by Dr Paul Davies, a mathematical physicist at King's College, London.

The title of this astonishing book is perhaps ambiguous. By "other worlds," Dr Davies does not mean Mars or Jupiter, or even the so far unknown worlds which are likely to be orbiting distant stars. He means that according to modern interpretations of quantum theory, our universe, with all its myriad stars, is but one of an infinite number of other universes which co-exist, parallel in time, in a domain which scientists call "Superspace."

This idea of an infinite number of parallel worlds was first put forward in 1957 by the Princeton physicist Hugh Everett, who described it as a "metatheory," meaning something that is half theory and half speculation. But, as Dr Davies reports, experimental evidence has now done much to remove the "meta."

Obviously, this cannot be proved either by abstract thought or by gazing into the night sky with binoculars. It arises from the observed behaviour of particles. Heisenberg found by experiment in 1927 that electrons, which orbit the nuclei of atoms, behave in a way which defies common sense.

Complete information about *both* the positions and the speed of electrons is unattainable. The more accurately we know either one of them, the less we know of the other. It is not that our instruments are inadequate for the task; rather, *the knowledge does not exist.* And so over the entire structure of matter (and of space and time, which cannot exist in the absence of matter) there is a vast uncertainty. Even to the eyes of an omniscient being, our universe would only be an approximation. Not even God can possess non-existent information.

Reality itself fluctuates. In the mini-micro world explored by quantum theory, which involves distances 100 billion billion times smaller than the diameter of the atomic nucleus, the structure of matter seems almost to come alive, with a foam-like character of twisting bridges and looping "worm-holes," a chaos of continuous change.

In the larger universe, the collapse of a distant star into a black hole releases "gravitational waves" which minutely affect the fabric of space and time through which the earth moves in its orbit. As Dr Davies points out, "the very space we inhabit takes on the features of a quivering jelly."

He describes a laboratory experiment involving two light waves that are simultaneously sent through two polarised panes of glass. It appears that each light-wave somehow *knows* in advance what the other is going to do, and itself does the same. If only one universe existed, there would be no way for this "knowledge" to be transmitted, and the experiment is said to show that another universe is interacting with our own.

All the "other worlds" are said to be branching from each other in an infinite tree at every micro-second of time. There will be one world in which Gerald Ford won the 1976 election, and a third where all the continents of the world were different and America never existed.

The one thing we cannot do is to travel between one such world and another—yet.

A message into the past

In vain men tell us time can alter
Old loves or make old memories falter.
                    —Swinburne, "Age and Song"

WINSTON CHURCHILL once remarked that the tragedy of Hitler's war was that it was avoidable. But a stronger point of view is receiving growing support from theoretical physicists, namely that the last war *is still avoidable.*

Let us see what they mean by this statement, which at first sight must seem even madder than the theory of Dr Paul Davies, which I reported last fortnight, of the existence of an infinite number of worlds, parallel in time to our own, where everything that can happen does happen.

We can now take this matter still further. It is proposed that by currently known science it may one day be possible to construct a Time Machine. This would be nothing less than a device through which one could send a message into the past—a message that in this case would presumably inform someone of the desirability of assassinating Hitler *before* he became dictator of Germany.

A progress report on speculations into the feasibility of time travel is given in a recent issue of the science magazine OMNI by Dr Robert L. Forward, chief scientist at the Hughes Research Laboratories at Malibu, California.

Dr Forward reminds us of the prediction made in 1963 by the New Zealand scientist Roy Kerr that a spaceship could pass unscathed through a fast rotating black hole, and, even more important, of the 1974 calculation of Prof. Frank J. Tipler, of the University of Maryland, that a black hole rotating in space is nothing more than a natural Time Machine.

Let me explain this. According to Einstein's general theory of relativity, the fabric of space and time are warped by the presence of a large mass. A rotating black hole with a mass several times greater than the sun's would create a "whirlpool" of space and time, through which a spaceship could travel almost instantaneously into another region of space.

But the ship can only make this journey through space if it simultaneously travels *backwards* in time. From this, Prof. Tipler concluded that any fast-rotating massive object must be a time machine.

Now using artificially-constructed black holes for journeys through space or time is a project that is probably centuries ahead. To construct a black hole sufficiently large for human passage would today cost about two thirds of all the wealth in the world.

But it might be economical, within the next century or so, to construct a small black hole merely to send *messages* into the past. How? By some advanced form of engineering, it should be possible to compress a medium-sized asteroid of about 15,000 million tons to the size of an atomic nucleus. It would then be a mini-black hole, and any signal passing through it, provided it had an extremely short wave-length, would travel backwards in time.

Yet is there not a fundamental objection to the idea of sending messages from the present into the past? It might result in altering the past, which is surely unalterable.

This objection, however, does not apply, if Dr Davies's theory of infinite parallel worlds is correct. *The past world which receives our message would thereupon cease to be the same world which it was before receiving the message.*

The past would have been altered by receipt of the
· message. The alteration might be imperceptible—if the message were ignored—or it might be very great, if it

caused someone to assassinate Hitler. In the latter case, we might find ourselves living in a world where the Second World War had never taken place. Our history books would no longer contain any record of this non-existent war. Reality itself would have changed.

The argument for a Time Machine has only one slight flaw. Why, if backwards time travel is scientifically feasible, have we received no visitors or messages from the future? One might answer that perhaps we have and that we are unaware of them. We follow the argument round and round in circles, always on the point of proving the case for time travel but never quite succeeding. Whether true or false, it has a tantalising fascination.

Newton in his orbit

THE appearance of Sir Isaac Newton on the back of the new pound notes reading a copy of his "Principia Mathematica," with his reflecting telescope, his prism, and a diagram of the solar system, reminds us of how much Einstein, of whom so much has been written, owed to the great English discoverer.

Einstein generously acknowledged his debt to Newton's mathematics after he had formulated his relativity theories. "Newton, forgive me," he wrote. "You found the only way which, in your age, was possible for a man of the highest thought and creative power."

Yet very different were the careers and characters of these two great pioneers of our exploration of the universe. Einstein liked to popularise his ideas. Newton loathed popularisation, and in fear of criticism laboured to make his discoveries as obscure as possible.

A Nazi professor once called a public meeting to

denounce the relativity theory. Newton would have shunned such a meeting in mortification. Einstein went along to heckle.

One result of Newton's intense shyness was that the average person still probably knows less about his life's work 250 years after his death than is known of Einstein's 24 years after his. By the time Newton died in 1727, his three laws of gravitation had been studied only by a small circle of scientists, while at Einstein's death the rudiments of relativity were known to millions.

Why did Newton's system of astronomy remain mysterious for so long? Only today, when rockets and jet aircraft make use of his famous Third Law, "that every action has its equal and opposite reaction," do we begin to appreciate it.

The reason must lie in the characters of the first Fellows of the Royal Society, of which Newton was a leading member. Seldom was there a group of people at once so brilliant and so mutually suspicious and childish in their personal quarrels.

Newton and Robert Hooke, the pioneer of the science of elasticity, hated each other because Hooke believed Newton had stolen his data on light and colour. Hooke detested nearly everyone because he suspected them of knowing that he had once made a living as a waiter.

John Flamsteed, the first Astronomer Royal, spoke rancorously about Edmund Halley, of comet fame, whom he had never forgiven for pointing out an error in his tide-tables. Newton and Halley treated Flamsteed like a flunkey, besieging the new observatory at Greenwich, demanding that Flamsteed publish the star-chart which he was painstakingly compiling.

The perfectionist Flamsteed petulantly refused, boasting of the pleasure he had taken in keeping Newton waiting outside his rooms. Halley and Newton at length seized the half-completed star-chart and published it without Flamsteed's permission.

Flamsteed in his rage destroyed as many copies of the chart as he could lay hands on. Newton, he declared was a "puppy," and Halley a "lazy and malicious thief." Newton, typically, tried to withdraw from the affair by putting all the blame for it on to Halley.

Newton was so nervous about the disputes which he sometimes helped to instigate, and the probable reaction of his "colleagues," that he only agreed to allow the Royal Society to publish his laws of gravitation in 1687, more than 20 years after he had worked them out.

When his manuscript was laid before the fathers of science, there was a furious reaction from Robert Hooke. He claimed in his usual truculent way that he, and not Newton, had worked out the movements of the heavenly bodies, and that he had explained them to Newton in writing six years before.

After a long dispute in which Hooke was in turn attacked by Sir Christopher Wren, the exasperated Newton agreed, much against his nature, to include a terse passage giving some slight credit to Hooke and others. But it was too late. The president of the society, Samuel Pepys, the diarist, had lost his nerve during this row, and, fearing a further controversy, backed out of the agreement to publish Newton's work. It was published soon afterwards at the private expense of Edmund Halley.

The guts of Newton's "Philosophiae Naturalis Principia Mathematica" ("Mathematical Principals of Natural Philosophy") amounted to three famous laws:

1. A body remains in motion at a constant velocity unless acted on by a force.

2. The attraction between two bodies in space increases as the square of the distance between them lessens. This is known as the inverse square law and applies in the same way as magnetic attraction.

3. This law I have already mentioned.

The certainty of uncertainty

IN THE celebrations of Albert Einstein's centenary year, little has been said of the late Werner Heisenberg, the only man to have contradicted Einstein and be proved right, and who laid the cornerstone for a new science as important as relativity itself.

"Heisenberg may have been here," says a joke which sums up quantum mechanics, which even today seems as mysterious as the personality of its founder. Heisenberg's famous principle of uncertainty states that certain information is for ever unobtainable. This is not due to any deficiency of information-gathering, but rather to the nature of knowledge itself. Consider this question: *Could an infinitely powerful computer make an infinitely accurate prediction?*

Since an infinitely powerful computer cannot exist, the question might at first sight seem merely philosophical. But the question is whether only the absence of supremely sophisticated instruments hides from us the totality of knowledge.

The Marquis of Laplace, in the early 19th century, suggested that God must be infinitely well-informed, even if man was not. A supreme intelligence, he said, "which knew at a given instant all of the forces by which nature is animated, and the relative positions of all the objects, could include in one formula the movements of the massive objects in the universe and those of the lightest atom. Nothing would be uncertain to it. The future, as the past, would be present to its eyes."

Werner Heisenberg, in his great flash of insight of 1927, afterwards confirmed by experiment, showed that Laplace was wrong and that the answer to the question was "no." Not even God could possess non-existent information.

His great uncertainty principle showed that electrons, the particles that orbit the nuclei of atoms, can never be located with absolute accuracy. The more precisely we define their positions, the less we know of their speeds. And the more we know of their speeds, the less we know of their positions.

This is not just a question of the information being unobtainable because of inadequate instruments. The complete information does not exist, since the behaviour of the electron is itself altered by the presence of the observer. In short, the popular expression "God knows!" may at times be hopelessly optimistic.

Heisenberg's discovery was a shock to many people, since the supposed principle of cause and effect had been an anchor of science since the days of the ancient Greeks. Einstein himself was profoundly disturbed by it, protesting that "God does not play the dice." "You can't tell God what to do," answered his colleague Neils Bohr. This year the cosmologist Stephen Hawking took the paradox even further: "Not only does God play at dice; He sometimes throws the dice where they can't be seen."

The certainty principle, so far as we can tell, applies to the micro-system of electrons rather than the larger universe of planets, stars and galaxies.

Yet this prohibition is not absolute. The moon, according to Isaac Newton's first law, will continue to orbit the earth in some trajectory or other until the end of time, unless acted on by a force. The orbit will change as the earth and moon exchange tidal drag, but it will be an orbit whose path can be predicted with almost absolute accuracy.

Almost absolute. That is the point. The moon just might, for no apparent reason, cease its orbit and crash into the earth. The odds against this happening during a period equal to the lifespan of the universe are so remote as to involve a number almost beyond possibility of calculation, but they are not zero.

It is amusing to speculate whether the uncertainty principle could apply also to the human personality. Like the electron of which we seek in vain to know two things, its position and its' speed, the human personality has two distinct sides, the intellectual and the emotional. Our information about one side cancels out our information about the other. We may know a great deal about a particular person's intellect, or about their emotions. We may know everything either about the one or the other, but we cannot, it seems, know everything about both.

The reason appears plain. When people are thinking sufficiently deeply, their emotions are hidden. When in a strongly emotional state, their capacity for rational thought appears correspondingly to have diminished.

Since a person's mental state is usually the result of external events or other people's behaviour, we can say that the personality, like the electron, is itself altered by the presence of the observer.

Could this uncertainty explain our fascination for the human mind? Could the unpredictability of human nature explain our love of art, because certain things are both unknown and unknowable? I offer the idea for what it is worth.

PSEUDO-SCIENCE

Science produces more than its fair share of char-latanry. The claims of astrology are only surpassed in improbability by belief in spoon-bending and the supposed dangers to shipping in the Bermuda Triangle.

The rubbish of astrology

PSEUDO-SCIENCE is with us again. We have had solemn "proof" that astrology is a true guide to human personality, and that our natal stars rule our lives. We are told of a "new scientific breakthrough," and that "science is emphatically changing its attitude" towards horoscopes.

These findings have been accompanied by special newspaper "pull-out" sections, and preceded by costly television advertising which included views of a radio telescope and one ludicrous shot of a man in a white coat feeding a Zodiacal chart into a computer.

The new evidence, for what it is worth, is that various groups of quite respectable sociologists and statisticians have charted the birthdays and hours of birth of tens of thousands of people in various professions. An overwhelming number have been found in the actual professions most suited to the temperaments supposedly bequeathed to them by the star constellations under which they were born. Scorpio, for example, the astrological sign for men of action, has claimed an unexpected number of British and American Army officers.

The findings have been gravely considered by several people who should know better. Among them is Dr Hans Eysenck, the distinguished Professor of Psychology at London University who, while admitting that much of astrology is fraudulent pretence, says that the new findings are "disturbing and uncomfortable."

Now a little bit of *astronomical* knowledge, as opposed to astrological, shows us that these findings, along with the rest of astrology, are demonstrable rubbish.

I must explain. In about 100 BC, the Greek astronomer Hipparchus noted that the sun, as seen from Earth, followed a circular path through the heavens that took it through 12 constellations of the stars, remaining approximately the same amount of time, about 30 days, in each. These constellations have, of course, no physical significance. The stars that comprise them are not especially near to each other, they have little in common and their apparent groupings are arbitrary assumptions of the Earth-based human eye.

A century after Hipparchus, the fortune-teller Thrasyllus, a lickspittle courtier to the superstitious Roman emperor Tiberius, constructed an elaborate system for prediction based on the sun's position in these constellations, or signs of the Zodiac, at a person's birth.

The Romans of the time, just like astrologers today, believed everything he said. "Thrasyllus is always right," said Tiberius's mother, the empress Livia. The very modern word "disaster" is based on an ancient Greek word meaning "evil star."

But even while Thrasyllus was making a fortune by telling fortunes, something was going horribly wrong. The universe was just not behaving as it should. The Earth, as Copernicus discovered 1,400 years later, was continually wobbling in its orbit, a phenomenon which we call the precession of the equinoxes. The result is that the times of the sun's passing through the constellations of the Zodiac are today vastly different from what they were in classical times. The fact is ignored by astrologers.

The astrologers say, falsely, that the sun enters Capricorn on Dec. 22 each year, spends 30 days in it, and then spends about the same time in each of the other 12 constellations before returning again to Capricorn.

In fact, the Sun spends the following number of days in each constellation, starting with Capricorn on

January 16: Capricorn, 30; Aquarius, 24; Pisces, 39; Aries, 22; Taurus, 35; Gemini, 26; Cancer, 21; Leo, 38; Virgo, 47; Libra, 25; Scorpio, 6; Ophiuchus, 18; and Sagittarius, 34.

Ophiuchus? Where did he come from? This new 13th constellation of the Zodiac was unknown to the ancients. Astrologers have nothing to say about it. Any statistician can see that a large number of those brave British and American soldiers cannot have been born in Scorpio at all, but in mysterious Ophiuchus!

So much for the Zodiac. We are told also that the position of particular planets in the sky at the time of people's birth strongly influences their characters. The astrologers point to Hitler and other Nazis, born when Jupiter was high in the sky, which supposedly enhanced their "expansionist" traits.

Hitler's birth at about 6.30 p.m. on April 20, 1889, indeed occurred when Jupiter was high. But so what? We have only to consult *The Observatory* magazine for 1889 (Vol. 12) to learn that Mars and the love-planet Venus were also in the sky at Hitler's birth. We can guess what the astrologers would have said if Hitler had been a famous lover instead of a mass murderer.

A planet is said to have powerful influence if it is just over the horizon when a person is born. But a celestial object radiates *least* when just over the horizon because of obscuration by the Earth's atmosphere. Once again, a major astrological prediction is seen as not just unlikely, but physically impossible.

Exposing the myths

A REMARKABLE new journal has appeared. It is called *The Zetetic,* meaning "the sceptic" in Greek, and it is written by scientists specifically to investigate

squadrons of flying saucers.

Then he "discovers" Uri Geller and writes the best-selling book "Uri" which relates countless marvels. Geller is delighted and confirms every word.

But there is a group of tiresome fellows who insist that every word of it is a lie, and that Geller's performances are fraudulent. They can do all his tricks by sleight-of-hand. These are the professional magicians who resent prostitution of their art.

The most active of them is the great Canadian conjuror James Randi. He is untiring in his efforts to expose Uri Geller. Since Geller won't perform in the presence of conjurors ("They're out to get me," he complains), Randi often has to pretend to be a reporter.

In this guise, in 1973, he enjoys himself hugely in the New York offices of *Time* magazine. Geller does some clever tricks, but Randi sees through all of them. Puharich is furious and alleges that Geller has been "lynched by a kangaroo court."

The same year a sad episode occurs at the Stanford Research Institute in California. Two quite distinguished scientists, Russell Targ and Harold Puthoff, decide to examine the "Geller effect." They little dream that anyone can deceive them. After all, they are scientists!

They perform tests with Geller which appear to prove the powers of parapsychology. The wool has been pulled neatly over the eyes of these two great men, and their report is published in the prestigious journal *Nature*. Geller's fame seems assured.

But things go wrong. His manager Yasha Katz, who arranged many of his hoaxes, goes on Italian television to relate how they were done. Geller has apparently neglected to pay him. Millions of people learn, one assumes authoritatively, that Geller's psychic power has been faked.

Imaginations in hyperspace

At the Commonwealth Institute in London, I witnessed a pitiful display of public credulity. Four hundred apparently well-educated people had turned up willing to believe everything they heard at a seminar on "the frontiers of physics and the paranormal."

A certain Dr Andrija Puharich, who claimed to be a scientist in medical electronics, announced from the platform:

"I have witnessed experiments in which young men have travelled into the past and into the future, and then back into the present. I have seen people travel through hyperspace from one place to another without passing any point in between.

"Once Uri Geller," (the spoon-bending conjuror) "teleported himself through my window, having travelled through hyperspace from a point 36 miles away. And I personally bear witness that these events really happened."

It might be supposed that on hearing this stuff that the audience would roar with laughter.

But so widespread has public belief in such pseudo-sciences as astrology and the "paranormal" now become that people are apparently prepared to believe any wild story if the speaker can claim to be a "scientist."

Even more bizarre were the assertions of Tom Bearden, who was advertised as a "research scientist in NASA"—although he himself does not claim to have worked for NASA. He had his own interpretation of the 1958 Russian nuclear accident in the Urals which is supposed to have killed hundreds of people.

Even the unlikely explanation given by Zhores Medvedev in the New Scientist that the accident was somehow caused by the explosion of nuclear wastes was far too tame for Tom Bearden.

Bearden stated that the Russians had been experimenting with the feasibility of a "hyperspace howitzer." The idea was that a hydrogen bomb would be sent from Russia through a tunnel outside time and space and it would re-emerge and explode over an American city. This idea, he assured us, was "solid theoretical physics, not mysticism."

Let's sample a little more of Bearden's "solid theoretical physics." It is possible, he says, that a gigantic and terrible animal may come into existence. Already this creature was beginning to wake and "gnash its teeth."

This animal, which would be totally insane, would be brought into being by the collective subconscious of 4,000 million people, the whole world's population, by the process of "psychokinesis."

Psychokinesis is a popular idea among mystics, and was frighteningly portrayed in the recent horror film "Carrie." If I have understood the concept correctly, it means that if you think about something hard enough it will really happen.

Bearden's animal, he told us gravely, was none other than the famous Beast, whose existence was first postulated in the Book of Revelation.

Now a journey through hyperspace, although ludicrous in the context in which the lecturers described it, does have a real meaning in physics. It would be possible in theory for a spaceship to travel through a black hole and re-emerge at a distant point without passing any point in between.

But because black holes are objects of incredibly high density, one may calculate that if Uri Geller was to make such a journey on Earth he would need to go through a black hole 82,000 times more massive than the Earth itself.

The gravitational field of such an object would of course rapidly devour the Earth and everybody on it.

Perhaps this object is the terrible insane Beast which Mr Bearden fears.

Now as Arthur C. Clarke once remarked, one cannot expect to build a democracy from people who are prepared to accept statements like these without demanding proof.

Perhaps the most effective weapon against pseudo-scientists is ridicule. Few extremists can survive being laughed at. Ridicule once nearly destroyed Sir Oswald Mosley when he was at the height of his prestige in the 'thirties.

He walked solemnly to the platform to start one of his mass meetings. Drums rolled. Spotlights played. The tension was electric. He raised his arm in a Fascist salute. Amid the awful silence a faint cockney voice was heard: "All right, Oswald, you can leave the room!"

The audience collapsed in laughter, and Mosley's credibility was almost ruined. I would dearly love to see someone play a similar trick on our modern nonsense-pedlars.

ENERGY

Arguments about energy policy can produce strong passions, and nuclear power has specially vociferous opponents. I also discuss the threat to the global environment from burning too much oil and coal, the possibility that Russia might start a war to acquire more oil—and the tantalising notion of a perpetual motion machine.

The furious opponents of nuclear power

THE case for a large expansion of our nuclear power programme is now unanswerable. The relative economics are explicit. Nuclear-generated electricity costs 0·76 pence per kilowatt per hour, compared with 1·23p for coal and 1·42p for oil. These are the latest available figures, taken before the recent O P E C rises.

And yet there remains a mystery. Why, in view of these statistics, available to anyone who cares to telephone the Central Electricity Generating Board, and of the magnificent safety record of nuclear power in the West, do so many otherwise intelligent people oppose it with such virulence?

Apart from people with obvious cynical motives, like Arthur Scargill, who would like to suppress any activity that would reduce the impact of a coal miners' strike, and the Communists, hoping that Britain will one day depend for her energy on Soviet goodwill, there are many, without any reasonable cause, who look on the peaceful use of the atom with dread.

These feelings can easily explode into hysteria and personal vilification against pro-nuclear supporters. The conduct of anti-nuclear protesters is sometimes reminiscent of that of Marat, the great French revolutionary, who, "at the slightest opposition would show signs of fury, and if one persisted in one's opinion would fly into a rage and foam at the mouth." Consider, in this context, the strange story of Dr Inhaber and Prof. Holdren.

The mild-mannered Dr Herbert Inhaber is a scientific adviser to the Atomic Energy Control Board of Canada. In 1978, he published a modest report, drawn from existing literature and accompanied by an official preface from his employers, in which he attempted to assess the relative risks of accidental death from nuclear power and rival energy systems.

His main conclusion, which Lord Rothschild included in his famous B B C lecture on "Risk," was reassuring:

Estimated range of deaths for a specified energy output (10,000 megawatts):

Coal	50-1,600
Oil	20-1,400
Wind	230-700
Solar space heating	90-100
Nuclear	$2\frac{1}{2}$-15
Natural gas	1-4

Comment from other scientists was at first favourable. There were criticisms and suggestions for improvement. Dr Inhaber made various corrections, and the report ran through four separate revisions.

At length it fell into the hands of John P. Holdren, a professor of energy and resources at the University of California, and a passionate opponent of nuclear power.

Prof. Holdren began to rage. He denounced the report as "the most incompetent technical document I have ever known to have been distributed by grown-ups." In a further outburst he called it "the shabbiest hodge-podge of misreadings, misrepresentations and preposterous calculational errors I have ever seen between glossy covers."

His press officer, Mr Kent Anderson, was incensed by a remark in the official preface to the Inhaber report that it was useful "food for thought." Using the report as food for thought, retorted Mr Anderson, was "equivalent to eating garbage."

The professor then launched into a massive propaganda campaign. He wrote twice to the Canadian Atomic Energy Board, demanding that they repudiate Dr Inhaber, who was a "serious embarrassment" to them. They refused to do so.

He bombarded the scientific press with personal

attacks on Dr Inhaber. His most serious charge, of which he has yet to produce evidence, was that Dr Inhaber's errors, which he admitted and corrected, were beyond the realm of honest mistakes and showed a conscious attempt to mislead.

When Dr Inhaber was revising his report, he was "manipulating the figures." In other words, according to Prof. Holdren and his friends, when Dr Inhaber says that nuclear energy is safer than all energy systems except for natural gas, he is not merely wrong; he is deliberately lying.

The professor admits that the tone of his campaign is not the usual one employed by dissenting scientists. But he refuses to concede, as reasonable people might think, that it is also grossly improper. He glories in it. He and his friends hope their vehemence will persuade laymen that their complaint against Dr Inhaber is "not just a dispute among experts." One cannot help agreeing. Nothing in Prof. Holdren's statements suggests expertise in anything, except for ungrammatical invective.

Why this fanatical opposition to an essential and proven technology? What is the emotional pathway that can make a rational person hysterical at the mere sight of a fission reactor?

One used to think that the root cause of opposition to nuclear power was the memory of the atomic attacks on Japan in 1945, and a fear that nuclear power stations might themselves explode like atomic bombs, a thing which is in fact physically impossible, but which can arouse uneasiness by mere association with the word "nuclear."

This explanation appears somehow far-fetched. It might well be true in Japan, but surely not in countries thousands of miles from Nagasaki and Hiroshima. There might be a deeper reason, lying not in the word "nuclear" but in the word "atom."

For 22 centuries, from the time of the Greek philosopher Democritus to the discoveries starting in the 1920s of hundreds of sub-atomic particles, atoms were understood to be the bedrock of matter. The very word "atom" meant "indivisible" in Greek.

So by splitting the atom, we are doing something regarded as impossible during 60 generations. We are "playing God." It may be that root of the opposition lies not so much in fears that nuclear power may be dangerous but rather that it is blasphemous.

Nuclear power cannot of course produce everything. It cannot drive our cars and lorries, or give us chemical feedstocks or fats and dyes. For these we shall always need coal products, a fact which makes Arthur Scargill's stance seem ill-informed as well as political. Only in large-scale electricity generation will nuclear power be essential.

Everyone is anxious that it should be made as safe as possible. But the extreme environmentalists argue that because it can never be made absolutely safe, then it should be abandoned, even if this means that tens of millions will be unemployed through lack of electricity.

By this argument, every human activity should be abandoned. Eating and breathing pose innumerable hazards. Perhaps they should be stopped also.

Windmills of the mind

I WAS recently the target of a finger-wagging lecture from Prof. Sir Martin Ryle, the famous radio astronomer, on the alleged evils of nuclear power. Sir Martin feels that nuclear power stations are potentially so dangerous and take so long to construct that instead we should build windmills in large numbers.

Sir Martin's knowledge of the remote parts of the universe is so profound that I would not argue with him about it. But on matters of terrestrial energy, this learned man seems on much less certain ground.

He tells us, for example, how in 1966, a metal plate broke loose in the Enrico Fermi nuclear reactor outside Detroit, and that the accident "led to emergency plans to evacuate Detroit 48 hours before the emergency was over."

This is technically true, but is a wholly misleading statement. If the reactor had lost its coolant (which it didn't), the emergency core cooling system would have supplied it.

If the emergency cooling system had failed, the containment building would have contained the radio activity. And if it hadn't (although it is hard to see why), the radio activity would have dispersed harmlessly in the atmosphere.

And if the radioactivity was prevented from dispersing by a temperature inversion, a strong wind in one unlikely direction would have been needed to blow it all the 30 miles to Detroit.

For Sir Martin to use this minor incident, which hurt nobody, as the central part of his argument that nuclear power is excessively dangerous shows that he simply does not understand the technology which he is criticising.

Nuclear power is not absolutely safe. No large-scale energy generating system ever can be. But a well-constructed nuclear reactor has so many back-up systems, so many emergency-within-emergency procedures that go into action if the next one fails, that it is a far safer system than any other.

Sir Martin's enthusiasm for windmills I can only call eccentric. Sir Fred Hoyle calculates that 20 million windmills would be needed to electrify Britain, with 240 of them per square mile if they were land based.

But Sir Martin says that Sir Fred's figure is "absurd," and that one per square mile would be enough, with the rest apparently to be built on off-shore rafts.

The land-based windmills would be no bother, he

assures us. They would be no worse than pylons.

But is this so? Pylon wires do no worse than make a loud humming noise if you happen to be standing beneath them in a strong wind. But windmills? Let Sir Fred Hoyle, who is something of an expert on the subject, describe what they would be like:

"When in full operation such an ensemble of mills would make an appalling roar, and the number of serious accidents they would cause would run into hundreds of thousands each year." (*Energy or Extinction,* 1977.)

A horrible tragedy in New York gives some idea of what such a "serious accident" might be like. A helicopter on the roof of the Pan Am Building fell on its side. A whirling blade fell to the street below, and cut a girl in half.

Let us dispense with windmills. The power source of the stars is nuclear, of a more advanced form than we have yet achieved. It would be more appropriate if the astronomer Sir Martin urged us to strive towards attaining that power, rather than turning back to primitive epochs.

Coal-mining threatens the world

NEARLY EVERY year during the miners' wage negotiations, we get the seasonal platitudes from the coal industry: "coal has a great future;" "it is vital to attract more men into the industry," and "60,000 miners condemn nuclear power."

I must interrupt these rosy visions with some harsh scientific facts. A team of 15 American specialists has recently ended a two-year study with some very unpleasant predictions about what coal-burning could eventually do to the human race. Their detailed report to the US National Academy of Sciences succinctly explains the problem:

When coal is burned in a power station, carbon dioxide is released into the atmosphere; and there must be a limit to the amount of carbon dioxide which the atmosphere can hold before the world's climate begins to change radically.

The atmosphere contains a huge amount of carbon dioxide which has been produced by volcanoes and by various chemical reactions. But the burning of coal and oil, the earth's "fossil fuels," in power stations and elsewhere, is adding to the mass of atmospheric carbon dioxide at the rate of about 0.7 per cent each year.

Reserves of oil, as a major energy source, will obviously last a much shorter time than coal deposits, and so it is unlikely that oil burning will have the chance to do any of the irreparable damage that coal can do to our planet.

Carbon dioxide is not of course a poison like carbon monoxide. But its danger lies in this: while it is almost transparent to sunlight it absorbs heat. It therefore creates a "runaway greenhouse effect."

This situation exists in the most extreme known form on the planet Venus, whose atmosphere is mostly carbon dioxide, where the average surface temperature is a staggering 900 deg. F.

The present 0.7 per cent annual increase of atmospheric carbon dioxide back here on earth may seem very little (it is actually about 20,000 million tons a year), but if continued at this rate it will double within 100 years.

But it cannot continue at this rate. As economic growth continues, and as unimaginative leaders like President Carter embrace coal and reject nuclear power, the rate of increase will itself increase.

We may therefore reach a point within 50 or 60 years where the world's average temperature is increased by between 4 and 5 degrees F.

Only a madman would suppose that such an

increase in heat would have a negligible effect on our climate. The icecaps would eventually start to melt, and the ocean levels would rise with this new addition of water until our coastal cities were under water.

Dr J. Murray Mitchell, one of the contributors to the American report, sets out the choices we face:

"Suppose we ignore the problem until it is staring us in the face—perhaps in another 20 years—in the form of a clear signal that a global warming trend has begun that is unmistakably attributable to the further accumulation of carbon dioxide.

"If we delay action until then to phase over our principal energy sources from fossil fuels to other kinds of fuels on an orderly rather than on a crash basis, the transition will be likely to take another 40 or 50 years to complete. By then much of the damage will already have been done."

In a few years, one fifth of Britain's electric power will be supplied annually by nuclear power, and four-fifths by oil and coal. I would like to see our dependence on coal being reduced much more rapidly.

The menace of oil-starved Russia

THE Soviet Union is likely to run seriously short of oil in the next few years, a time when she may have become the world's leading military power.

By or before 1985, according to a consensus of Western economists and energy experts, the Russians may be forced to take some extreme form of action to increase their oil supplies, a prediction which must not exclude the possible seizure of the Iranian and Saudi Arabian oilfields. For they would be driven to robbery by an energy crisis born of technical incompetence.

Russia, now an exporter of oil, produces approximately 12 million barrels per day. This rate of production will have peaked either in 1980 or 1981, and will thereafter decline. Only by starting to cut oil exports from about 1982 onwards can Russia hope to maintain her present rate of economic growth.

Yet any reduction of the 1·5 million barrels per day which Russia now exports to her East European satellites would seriously depress living standards in those countries, provoking unrest which the Kremlin might be unable to handle.

Nor can the other half of Russian oil exports be cut without risking economic disaster. The 1·3 million barrels per day sold mostly to Western Europe brings in the foreign currency desperately needed to buy Western technology and food.

Surely, it will be argued, the shortfall could be made up by a massive effort to increase oil production. But in Russia oil-drilling techniques are incredibly backward. It can take a year for a Soviet crew to drill down to 10,000 feet, a job which an American crew can do in a month.

The Soviet oil industry, for example, lacks both the semi-submersible rigs necessary for drilling and exploring in Arctic seas and the seismic recording machines which must be used below 1,600 feet of permafrost, where new Siberian oil is most likely to be found. Indeed, one may speculate that had the Soviet Union been responsible for extracting oil from the North Sea hardly a drop of it would yet have reached the market.

Then why not increase production of coal, natural gas or nuclear power? After all, oil only supplies 43 per cent of Russia's energy, while she has four times more natural gas reserves than the United States, and more coal reserves than any other country.

Again, these options appear to be closed. Ninety per cent of the coal lies beyond the reach of present Soviet technology, and industry has no piping with which to carry natural gas thousands of miles from the icy fastness of Siberia, where the untapped reserves now lie. Nuclear power, again suffering from backward technology, only supplies 1 per cent of the nation's power, compared with about 13 per cent in both Britain and the United States. Nuclear fission may eventually save the Soviet economy. But increased nuclear energy will be a long-term solution only, since it takes between five and ten years to commission a nuclear power station.

Astonishing to relate, the Soviet Union does not even have a co-ordinating Energy Ministry, and there is evidence that until about two years ago the Kremlin did not expect a serious energy shortage.

Energy problems are managed by about 60 separate government departments. Chief among them are the Ministry of Coal Industry, the Ministry of Chemical and Petroleum Machine Building, and the Ministry of Construction of Petroleum and Gas Industry Enterprises—who report only to the Praesidium, which is headed by the ailing Prime Minister, Mr Kosygin*. This easy-going arrangement is hardly likely to lead to effective action in the short-term.

But the Kremlin leaders have to face the short term. What will they do in this situation? In a frightening analysis of these facts in the February (1980) issue of *Fortune*, Mr Herbert E. Meyer calculates that they will have three options:

1. Impose a harsh regime of austerity on their industrial workers, cutting wages and reducing temperatures in factories.

2. Buy oil in the open market.

3. Obtain oil in the open market without paying for it.

Option One, although the most desirable from the Western point of view, would be highly undesirable to the Soviet dictatorship.

For Option Two, there will simply not be enough available foreign currency. Russia could always raise money by selling from its secret reserves of gold, but here she would be trapped in an ever-increasing spiral

*Mr Kosygin has now been replaced by Nikolai Tikhonov.

of rising oil prices and falling gold prices. The more gold she sold, the less oil it would buy.

Option Three seems to be the surest way to preserve the Soviet regime.

Obviously, the choice of option Three would be intolerable to the West, and could very easily lead to nuclear war. For by about 1985 we will have reached a situation where both super-Powers will depend on Middle East oil, and there will only be enough of it there for one of them.

But present Soviet encroachments in the Gulf and the Horn of Africa suggest that Option Three is being actively considered. Only a truly massive Western military deterrence in the region can force a return to Option One.

Unfortunately, the West has so far done little to discourage Option Three. A mere 20 American warships deployed in the Indian Ocean and 1,800 Marines bobbing around in the sea will by themselves deter neither covert Soviet blackmail of the oil sheikhdoms nor a full-scale Russian military invasion. The American Administration assures us that 100,000 American troops could reach the Gulf within a fortnight of a Soviet move on the oilfields; but a fortnight is a long time in war.

Perpetual notions about motion

THE DREAM of building a perpetual motion machine, although such a device would violate the laws of physics, still attracts many enthusiastic would-be inventors. The United States Patent Office receives more than 100 applications a year from people who claim to have designed one.

The idea behind a perpetual motion machine needs precise definition. It would be a "closed system," a device which produces energy eternally without having any power source to drive it.

A simple example would be a self-filling vessel. A curved, open-ended tube is filled with water. Gravity forces the water round the tube so that it circulates in a never-ending flow.

Except that it doesn't. The water in this tube would simply stabilise in one position, and part of it would spill on to the floor. After that nothing whatever would happen.

I have not tried the experiment myself, so I am not sure in what way the water would stabilise. But stabilise it would, and there is no possibility whatsoever of obtaining perpetual motion from this device or from any other.

This is not only because of friction. The reason is enshrined in Lord Kelvin's Second Law of Thermodynamics which states, in effect, that all energy must eventually run down. The billiard ball must come to rest; the bicycle will move only when it is pedalled. The stars themselves will run out of nuclear fuel and turn black. If the universe collapses into a single giant black hole, space, time and matter will come to an end.

If this does not happen, the Cosmos will die when it runs out of heat, and all energy will be dissipated. This process will take uncountable billions of years, but the period will be finite and the end inevitable.

It might be supposed that the rush of a planet in its orbit through space creates a perpetual motion machine. Imagine—and this is only fantasy—that we were somehow to place an enormous "moon-mill" in the path of the moon's orbit.

Every 27 days the moon would strike one of its gigantic blades as it passed by. The moon-mill would turn and provide energy. Would not this be a perpetual motion machine?

It would not be, because, to a minute degree, the collisions with the moon-mill would slow the moon down. Energy would be transferred from the moon to the mill, and no new energy would be created.

The moon would have to lose this energy, and consequently alter its orbit very slightly, in order to obey Isaac Newton's First Law, which states that an object will continue its uniform motion "unless acted on by a force."

In the universe we live in—without taking into account any other possible universes—mass may be transformed into energy, but neither mass nor energy çan be created out of nothing.

Despite these obvious reflections, the United States Patent Office receives would-be perpetual motion discoverers with a surprising courtesy and patience. Mr William Feldman, Patent Office Deputy Commissioner, explained the procedure: "When we get these applications, we tell the applicant that a perpetual motion device is involved, and we offer to refund his filing fee of £33. "If he persists, after we have explained that his idea is based on something impossible, we may ask him to submit a working model, and that usually ends the matter."

But what if, despite its impossibility, the idea has some genuinely useful feature? Another official explained: "We ask the inventor to go back to the drawing board, eliminate the perpetual motion aspect by providing some energy input, and come back to us. A lot of good ideas are salvaged that way."

MISCELLANEOUS ITEMS

Tom Lehrer's humorous song about nuclear war leads to a miscellany of science articles which concludes this book. My favourite is the last one of all, on how we might be able to predict the behaviour of ghosts.

As the fallout settles . . .

PUBLICATION of a new book on the probable effects of a nuclear war, a lucid, competent, and appropriately horrifying study by the science writer C. Bruce Sibley, reminds us how completely our normal mental processes can break down when we contemplate this worst possible of catastrophes.

Many people become irrational when the subject is even raised. When, in 1961, Dr Herman Kahn published "On Thermonuclear War," analysing the effects of such a war in cool, mathematical tones, a reviewer in *Scientific American* lost his temper completely and denounced his book as "devilish and blasphemous." Kahn wrote with just indignation to the editor: "Sir, your journal is neither scientific nor American . . ." But the editor appeared just as angered as the reviewer had been by the book, and Kahn's letter was never published.

At the risk of being guilty of bad taste, I would go considerably further than Kahn. One way to learn to think uninhibitedly about a subject (which we must about this one if we want to survive it) is to make jokes about it. From Dr Strangelove onwards, the nuclear war joke industry has gradually blossomed, and I would like to show off some of the gems from my collection.

The song-writer Tom Lehrer announced in 1965 for the benefit of "war buffs" that if we wanted any good songs to come out of World War III "we had better start writing them now." His attempt at "pre-nostalgia" started like this:

Little Johnny Jones was a US pilot
And no shrinking violet was he.
He was mighty proud when World III was declared,

And he wasn't scared, not he.
And this is what he said, on his way to Arma . . .
gedd . . . on!
  "So long, Mom,
  I'm off to drop the Bomb,
  So don't wait up for me."
Pilots going off to "drop the bomb" reminds me of a
remarkable story I found in Peter Michelmore's biogra-
phy of Einstein. In 1922, the president of the
Japanese Diet in Tokyo told members that the distin-
guished Prof. Einstein was about to visit Japan.

But did anyone know what his theory of relativity
was all about? "Oh, yes, I know all about it,"
exclaimed one member. "It concerns the relationship
between man and woman."

Twenty-three years later there was a very loud bang
over Hiroshima, and it was learned that relativity
theory was about something quite different.

"Dr Strangelove" in 1965, which depicted a mad
general starting a nuclear war on his own initiative
"because the Commies are putting fluoride in our
water supplies," must have been one of the funniest
films ever made. Long remembered will be the slightly
less mad general in this film, played by George C.
Scott, who talked reassuringly about "acceptable
numbers of megadeaths." The President of the United
States, acted with great style by Peter Sellers, rebukes
the Russian Ambassador for resisting arrest when
caught taking illicit pictures: "You can't fight in here.
This is the War Room!"

One joke has been made about the neutron bomb,
that it is a "capitalist weapon" because it destroys
people but not buildings. But it is not generally known
that in some conditions the neutron bomb will kill the
poor and spare the rich.

The reason is obvious. Rich people have swimming
pools. All they have to do when they think a neutron
bomb is about to go off nearby is to dive in and stay
under. They could be safe for the crucial few seconds

because water absorbs neutrons.

But Lehrer, in an earlier song, uttered an extremely important inaccuracy, which I would like to correct:

We will all bake together when we bake,
There'll be nobody present at the wake;
With complete participation
In that grand incineration,
Nearly three billion hunks of well-done steak.

Several important studies have since confirmed that this pessimistic view is incorrect. There will not be "complete participation." In short, it is believed that at least half the human race would survive a full-scale nuclear exchange.

Herman Kahn, after his exhaustive researches, was the first scientist to realise this. After imagining "scenarios" of the worst nuclear disasters he could think of, he asked in each case: "Will the survivors envy the dead?" And the answer in every case was, "No."

For many years my favourite in the collection has been:

Hark to the tale of Frederick Worms,
Whose parents weren't on speaking terms;
So, when Fred wrote to Santa Claus,
He wrote in duplicate, because
One went to Dad and one to Mum—
Each asking for plutonium.
So Fred's father and his mother
(Without consulting one another)
Each sent a lump of largish size,
Intending it as a surprise.
These met in Frederick's stocking—
And laid waste some ten square miles of land.
Learn from this tale of nuclear fission
Not to mix science and superstition!

But I have never been able to trace the author.

Oppressed science in Russia

HERE is the story of more than a million people who live in an atmosphere of fraud, hypocrisy, corruption and terror. For those not recognising the description, this is the plight of most of the scientists of the Soviet Union.

The evidence comes to us from Mark Popovsky, a Soviet writer who published more than 20 books on science, but then fell foul of the authorities because of his accusations against Lysenko. Banned from publishing anything, he defected to the West in 1977, with a microfilm of the book he had been writing in secret for many years, an account of the suppression of scientific creativity in Russia. Entitled "Science in Chains" it was published by Collins in March 1980.

Even allowing for the fact that emigres have a tendency to see their former countries in the worst possible light, the picture that emerges from Popovsky's account is unrelievedly dismal. One wonders after reading it how the Russians have managed to invent anything at all since the Revolution.

Much of the trouble comes from the Soviet obsession with "bigness." The authorities believe that the bigger the number of scientists working on a problem, the more quickly it will be solved. This mania has resulted in the closing down of countless small laboratories, where creativity best flourishes, and the establishment of giant "institutes" or "science cities" where thousands of scientific workers spend their lives toiling on projects whose purpose they often do not know.

There are now 1,200,000 scientists in the Soviet Union, but the quality of Russian science seems to be in inverse proportion to the number of scientists. Most of them have to live in remote parts of the country, with low morale and an assembly-line mentality.

Few of them get any credit for their work. The director of the institute, who is usually not a scientist at all but a bureaucrat or a member of the KGB secret police, will often take credit for work done by a subordinate. If a paper is to be read at a conference abroad, to which its true author is not allowed to travel, the director will attend in his place and present the paper as his own.

A KGB colonel named O. Baroyan, expelled from the World Health Organisation for spying, obtained a "quieter post" as director of a Moscow research institute on epidemiology, microbiology and immunology. Having little knowledge of any of these subjects, and no scientific degree, he soon remedied these defects by ordering his subordinates to write his doctoral thesis for him.

"Having become a big boss," says Popovsky, "the former spy and saboteur Baroyan is now actively engaged in sabotaging science. The system that prevails in his institute is one of unmitigated terror. He bullies his subordinates on academic councils, dismisses anyone he doesn't care for, and treats his laboratory assistants with special ruthlessness."

When one of these assistants, a talented doctor named T. Kryukova, was reported to have considered resignation in protest against Baroyan's brutalities, he threatened in public: "I'll take her back into the institute if she applies to me this very day and kneels in my office. If she waits till tomorrow, she'll have to crawl to my office all the way from the lobby."

Secrecy reigns not only in military establishments but in all scientific institutes. Its purpose is not only to conceal the shortcomings of Soviet science, and the fact that most of it consists of copying goods purchased from the West, but also for personal gain.

The cynicism shown in the following story will

astonish people familiar with the open methods of Western science. In 1972, Popovsky wrote an article in *Pravda* describing the isolation of ischemic toxin, a substance that can prevent a successful regrafting of severed limbs and operations for heart disease.

The Minister of Health, Boris Petrovsky, was furious at the publication of the article. He telephoned the head of the laboratory concerned to protest. If the discovery was as important as the article had said, he declared, it ought to have been kept secret to avoid unnecessary excitement.

What was Petrovsky's real reason for wanting to keep secret a discovery that could save people from heart disease and make it possible to regraft severed hands and feet? It is very simple. He is Minister of Health. As Popovsky explains, "The way for a Minister of Health to ingratiate himself is to see that the Kremlin élite are well looked after when they need medical aid. If the discovery had remained secret, Petrovsky could have used it as he saw fit, restricting its benefits to the Party and the Government higher-ups, who would have been correspondingly grateful."

Nearly every Soviet scientist lives in fear. "I am not speaking now," says the author, "of the constant fears that beset the life of an over-talented scientist whose work attracts the attention of foreign colleagues, or the tribulations of one who is a Jew, or the panic of a laboratory head who finds that he has a dissident on his staff. People in these categories live in a state of unremitting terror.

"But even those who are not Jews or dissidents and have nothing on their conscience cannot feel at ease, for one can never know what the authorities will choose, at a given moment, to regard as criminal and subversive."

What is life like in the great "science cities"? At Protvino, he writes, "everything is calculated to induce the young scientist to sell his soul. Life is insufferably boring, all topics of conversation have long been exhausted, and people stopped visiting each other long ago." At Akademgorodok, near Novosibirsk, 35 per cent of the scientists habitually get drunk.

Genuinely creative scientists have the worst time of all. Longing to communicate with foreign colleagues, they cannot do so because of the vigilance of the authorities.

What is the point of it all? Why all this cruelty and waste of talent? The cause appears to be a mixture of arrogance, ignorance and stupidity, from the Politburo down to every party official, an attitude dating from Czarist times that scientists should be treated as virtual slaves.

Lenin expressed this with his usual crudeness in a letter to Gorky in 1919: "The intellectual forces of the workers and peasants are growing and gaining strength in the struggle to overthrow the bourgeoisie and its henchmen, the intellectual lackeys of capital, who imagine they are the brains of the nation. Actually, they are not the brains, but the shit."

How Plato obstructed knowledge

MR ROBERT O. ANDERSON, owner of a well-known British newspaper and the chairman of Atlantic Richfield Oil, has a habit, happily uncommon among industrialists, of requiring his executives to attend seminars, outside working hours, on the subject of Plato's philosophy.

"Requiring" does not mean that the obligation appears in their contracts of employment. It means rather that failure to attend these seminars implies "demonstrating a lack of seriousness," in the words of an Atlantic Richfield official, which would naturally inhibit their promotion.

But I cannot help feeling, however much Mr Anderson may admire Plato, that Plato would have regarded Mr Anderson with abhorrence. It is not hard to imagine the indignation with which he would have beheld oil-rigs and newspaper presses, and the fury with which he would have expelled such mercantile contraptions from his Republic.

Plato's "Republic," that authoritarian document in which he set down his vision of an ideal society, was for more than 1,500 years perhaps the greatest obstacle in the path of science and technological progress, and it still finds echoes among extreme environmentalists today.

Plato would have disliked everything that Mr Anderson does and stands for. He hated utility and he hated the notion that science might have any purpose other than being a vehicle for sharpening the wits of the cleverest philosophers.

Plato had perhaps the most penetrating mind of the classical era, and he used it, with considerable success, to retard the progress of discovery. It was mainly due to his influence that, until the philosophical revolution of Francis Bacon at the end of the 16th century, scientists and inventors tended to be either despised as mere mechanics or burned as magicians.

Plato was interested in astronomy, but in the opposite sense to ourselves. Knowledge of the heavenly bodies or their motions was of no interest to him.

We must get beyond these mechanistic speculations, he says. We must neglect them. We must develop a mystic astronomy which is as independent of the actual stars as geometrical truth is independent of the lines of a diagram. We must use astronomy only to contemplate those things which the pure intellect alone can perceive.

It is the same with mathematics. Plato had no time for that vulgar crowd of merchants and inventors who used arithmetic and geometry to trade and construct machines. The sole purpose of this and all other sciences was to lead men to eternal and abstract contemplation.

He finds no great merit in handwriting, because the ready accessibility of information teaches men to be idle. Books should be discarded as soon as they have been read, and men should seek to increase their intellects by feats of memory.

He had little sympathy for medicine. He has no objection to cures for people injured in accident or battle, but sufferers from degenerate diseases, being doubtless punished for their gluttony or lust, should be allowed to die. They are unfit for mental exercise, and they become giddy in the head and interrupt serious discussion.

Plato effectively sabotaged the beginnings of Western science. Even the modest inventions made in the centuries after his death were despised by contemporaries.

The Roman philosopher Seneca described the achievements of inventing shorthand and central heating as "drudgery for the lowest slaves." He defended Democritus from the disgraceful accusation of having built the first arch, and Anacharsis of the charge of having invented the potter's wheel.

Even Archimedes, the greatest engineer of classical times, who held a Roman fleet at bay for three years with his lethal mechanical inventions, regarded this work as being a mere amusement, a relaxation from the rigours of high-minded philosophy.

The schools of Athens, in the 700 years between the time of Socrates and Plato and their closure by the Emperor Justinian, not only contributed virtually nothing to man's knowledge of the earth or the universe, they induced an intense feeling of guilt in anyone else who tried to do so.

Their intellectual heirs were the mediaeval schoolmen, who spent countless generations asking questions which could never be answered: how many angels can stand on the point of a needle? Can we be certain that we are certain of nothing? Are all departures from right equally reprehensible? Is pain good or evil? Can a wise man be unhappy?

The purpose of Plato and his followers was to engage in dialectic, to strive in never ending arguments, and to *cut off* science from any useful avenue. In Macaulay's words, "they filled the world with long words and long beards, and they left it as ignorant as they found it."

Mr Anderson's executives may indeed learn to sharpen their wits on Plato's philosophy, but it is hardly the most suitable intellectual refreshment if he expects them to return to their desks afterwards and make profits.

Ptolemy rehabilitated

SCIENTISTS can be as narrow-minded as other scholars. Large quantities of obscure knowledge can give their possessor an immense self-importance. Common sense is swept away and other academic disciplines are dismissed as irrelevant.

A recent book about the classical astronomer Ptolemy is an example of this phenomenon. Dr Robert Newton, a physicist at Johns Hopkins University in Maryland, decided that Ptolemy, the great savant of the 2nd century AD and one of the heroes of scientific history, must be a fraud. Ptolemy's methods of reaching theory from observation so baffled Dr Newton that he decided on the basis of his own astronomical knowledge that they must be due to trickery.

He began his book, "The Crime of Claudius Ptolemy" (Johns Hopkins University Press, 1978), in the thunderous tones of Burke impeaching Warren Hastings: "This is the story of a scientific crime. It is a crime committed by a scientist against his fellow-scientists and scholars, a betrayal of the ethics and integrity of his profession that has for ever deprived mankind of fundamental information.''

Now Ptolemy did make one very serious mistake, which retarded science for 12 centuries. He concluded that the earth is the centre of the solar system, and that the sun and the planets all revolve around us.

This belief was universally held until the Rennaissance, when Copernicus had the genius to perceive that the solar system made no sense if you tried to think of it in this way. The movement of the planets and the fixed stars could mean only that the earth itself was in motion.

But Dr Newton is not complaining about innocent mistakes. Ptolemy, he says, was a shameless plagiarist and inventor of data. He falsified the lengths of the reigns of Babylonian kings to suit his theories, and he quoted from scholars who never existed.

His great star-chart, the Almagest, meaning ''the Greatest,'' could not possibly have been compiled by the methods he describes, says Dr Newton. ''Ptolemy developed certain theories, and discovered they were inconsistent with observations. Instead of abandoning the theories, he deliberately fabricated observations so that he could claim they proved his theories.''

In the past few weeks Ptolemy's reputation has been cleared, and in a manner that throws no credit at all on the scholarship of Dr Newton. Prof. Victor Thoren, of Indiana University and Prof. Owen Gingerich, of Harvard, have pointed out that the astronomers of Ptolemy's epoch simply did not think in a modern manner. They were not really interested in determining the nature of the universe. They preferred to argue about it. They would sharpen their intellects in battles of wits. Logic was everything, experiment was nothing. They wished to prove truth rather than discover it.

They were the spiritual disciples of Plato, who thought that books should be thrown away after being read, to heighten the powers of memory; and of Socrates, who expelled a youth from his logic class for suggesting that the best way to calculate the number of teeth in a horse was to open the horse's mouth and count them.

Only in the 16th and 17th centuries, when the value of purely experimental science had been proclaimed by Francis Bacon, when *all* random observations began to be scrupulously reported, did scientists begin to think as they do today.

Dr Newton seems unaware of this change. He does not seem to have read Macaulay's essay on Bacon, which explains more eloquently than other documents changes in scientific thought between the 19th century and the epoch of Ptolemy.

Ptolemy is accused of falsifying the dates of 12 lunar eclipses. This is a remarkable piece of mathematics by Dr Newton. Ptolemy says his predictions of the eclipses proved correct. But the chances of *each* of them being correct by these methods, says Dr Newton, was 1 in 10. Therefore the chances of *all* of them being correct was 1 in 10 to the 12th, or 1 in a trillion. A full account of this controversy can be found in the March (1979) issue of *Scientific American*.

Dr Newton applies the modern method of calculating the probability of dice-throws. If the chance of throwing a six is 1 in 6, the chances of throwing two successive sixes is 1 in 36 (six squared), and of three successive sixes 1 in 216, and so on.

But these calculations require knowledge of the number of sides to a dice. Dr Newton had no means of knowing the total number of lunar eclipses Ptolemy predicted, only of those he reported. So the 1 in 10 assumption is invalid. The complex behaviour of people living 18 centuries ago cannot be detected by gambling table arithmetic.

It is a pleasure to rehabilitate the great Ptolemy. He retarded our discovery of celestial mechanics, but he compiled the most extensive star catalogue of antiquity, and he discovered a means to predict the movement of the planets. He also did much to counter the hostility towards science of the early Christians.

Tinkering with the mind

WHETHER THE CIA were right or wrong to carry out mind-experiments to control human behaviour is a question for the moralists. But what seems far more interesting is that this was one of the first serious attempts to explore a fascinating branch of psychology.

The human mind is a far more powerful and subtle instrument than has yet been realised. To give an idealised example of what I mean, consider the fact that there are some 10,000 million neurons, or nerve centres, in each of our brains.

The total number of possible permutations, or re-arrangements, of these neurons is the factorial of 10,000 million; this means 10,000,000,000 multiplied by 9,999,999,999, then by 9,999,999,998, and so on down to 1.

The resulting number is so gigantic that we may think of it like this: if all the atoms in the universe were themselves universes, and all the new atoms in these new universes were also universes, we would then have a total number of atoms only 10 times greater than the factorial of 10 billion.

Now it is obvious that all but a tiny fraction of these neuronic re-arrangements would destroy the brain. But that tiny fraction remaining could conceivably produce untapped and scarcely imaginable mental powers.

The reason why some people are far more success-ful in life than others is partly because they really use their brains, perhaps not the fullest power, but to something approaching it by sheer concentration. "Genius," said Thomas Edison, the electrical pioneer, "is 1 per cent inspiration and 99 per cent perspiration."

Let me give some examples. Isaac Newton, having retired in 1666 to the solitude of his stepfather's farm at Woolsthorpe, concentrated for periods of up to 10 hours at a stretch on the problem of gravity.

Most people cannot concentrate for more than about one hour or two hours on a single problem. But Newton's powers of concentration were so great that after reading the description by his biographer E. N. da C. Andrade, one is almost surprised that his mind did not burn a hole in his writing desk.

Einstein, in Zurich in 1904, seemed totally cut off from this world as the special theory of relativity began drawing on him. A friend complained: *"Albert wanders around in a daze. When people speak to him he does not answer."*

While Joyce Wethered was sinking a long putt on the 11th green at Troon in the 1925 British Women's Golf Championship, a particularly noisy train roared past. "Weren't you put off by the train?", she was asked.

"What train?", she said in surprise.

The CIA had it in mind to play exotic tricks with other people's minds. Now it is fairly easy to implant an irrational obsession in someone else's mind, but the trick is to do it *undetectably* so that both the victim and his friends are unaware of the change.

It is difficult to know whether present-day cold war psychologists are having any success with such tricks, since they cannot publish their results in the scientific journals. But science fiction offers some intriguing possibilities.

In Richard Condon's 1959 novel "The Manchurian Candidate" a Korean War veteran comes home unaware that he has been captured and brainwashed and that anything is wrong with him. Then he receives an anonymous telephone call suggesting that he plays solitaire. This is the enemy's trigger message. His brain "flips over," and he will do whatever the caller tells him, and remember nothing afterwards.

Prof. Isaac Asimov's epic novel "The Foundation Trilogy" is even more elaborate. One of its characters is a dictator called the Mule, who is able to reach out with his mind and "adjust" people's emotions just as an ordinary person can adjust the dials on a clock.

He can thus "convert" his enemies into loving him. Armies cannot fight against him, and people cannot assassinate him, because he can detect the murderous intent in their minds.

He is opposed by a group of psychologists called the Second Foundation, whose truly terrifying power of deceit and mind manipulation the CIA would have dearly liked to possess.

Whether such exotic telepathy is really possible, the future will tell, for modern claims to have achieved it invariably prove bogus.

Some peculiar occurrences

AN unpleasant episode occurred in Bucharest on July 25, 1872. To everyone's horror, it was raining black worms. Their source remains unknown.

This incident is recorded in a fascinating new book, the "Handbook of Unusual Natural Phenomena," edited by William R. Corliss (The Sourcebook Project, Glen Arm, Maryland, 21057, USA) which documents thousands of peculiar natural events.

Mr Corliss describes only those events which were reported in reputable scientific journals, where all material is rigorously screened, and "misidentifications and hoaxes are kept to a minimum."

He gives us a feast of amazing tales. Consider the

strange story of the Barisal Guns. One night in 1871, a man was on the deck of a steamer in the mouth of the Ganges near the town of Barisal. The weather was clear and calm. Then, from far out to sea, came the boom as of distant cannon.

The sounds seemed to come from two different points, as if opposing fleets were exchanging fire. But no warships were present. Similar natural booms have been reported throughout the world. What is the explanation? Pockets of gas escaping from the ocean, the sonic booms of falling meteorites, rocks cracking under seismic pressure; all have been blamed, but the true cause is uncertain.

Ball lightning is perhaps the nearest thing we have to ghosts. There is no satisfactory explanation of it. It is so variable. It can be the size of a pea or as large as a house. It may be violet, red, yellow, or even transparent, and it can change colours during the few seconds of its existence.

It may glide silently or with apparent inquisitiveness round a room and then either quietly dematerialise or vanish with a violent explosion. It is usually spherical in shape, but rods, dumb-bells, spiked balls and other shapes have been reported.

An unusual ball of lightning materialised inside a screened patio in Dunnellon, Florida, on August 25, 1965. The temperature was in the nineties and the day was overcast with a slight drizzle.

A couple were sitting in aluminium chairs a few feet apart. The woman had just swatted a fly when a ball of lightning about the size of a football appeared in front of her.

The fly-swatter was "edged in fire." The woman dropped it in her fright, and the lightning ball vanished with a bang "like a shotgun blast." This incident was described the following year in *Science*. The cause of the apparition was plainly electrical, but it is impossible to be more specific than this. The physics of ball lightning is still almost wholly mysterious.

A thunderstorm in a clear sky may seem an even more unlikely event. Yet one occurred on the northern edge of the Gulf Stream on April 27, 1886. A ship's crew witnessed it as reported in *Scientific American.*

Lightning does odd things. One day during a thunderstorm in Kensington, New Hampshire, in 1856, it made a 30-feet deep hole in a field. In Forfarshire, Scotland, in 1867, lightning (one presumes it was lightning) penetrated a haystack and made a neat, round hole in the ground underneath. According to *Symons's Monthly Meteorological. Magazine,* a substance resembling lava was found in the hole.

Bedouins in Sinai speak of monks who were damned for eternity, whose Matin bells may still be heard in a place called Gebel Nakus. Scientists are not too sure about the monks, but they have heard the bells.

The Egyptian hero Memnon died in the Trojan war, and his mother Eos still mourns for him at sunrise at the temple of Karnak in ancient Thebes. So at least says the legend.

Scientists visiting Karnak have heard at sunrise in the temple a sound variously described as the ringing of a bell, a note of music or the snapping of a cord, all of which might be vaguely symbolic of a mother mourning for her son.

Some wind effect is of course responsible. A similar experience was reported in *Nature* in 1932 from Angmering-on-Sea, Sussex, where some people indoors heard a beautiful natural melody played in E major. The sound was eventually traced to wind rushing out through the overflow pipe of a bath.

Where then did those black worms come from? The most likely culprit, as Mr Corliss points out, is a tornado.

The voice of Arthur C. Clarke

A NEW VOLUME of essays by that great futurologist Arthur C. Clarke contains a number of pungent and witty observations about the medium and long-term technological future which deserve to be recorded.

In "The View from Serendip" (Gollancz, £5·50), the man who invented the telecommunications satellite and who taught thousands of Indian peasants to use such a satellite with nothing more than a TV set and a three-yard-wide, umbrella-shaped receiving antenna made of chicken wire, speculates about the ultimate frontiers of science.

Mr Clarke roams easily among the most astonishing propositions and anecdotes. Could we ever recapture sounds made in the past—long before there were any tape recorders? How did second-rate minds react to news of the invention of the camera? Did the poet Tennyson understand the theory of continental drift? What was the most idiotic remark made by a British Cabinet Minister in the last two years?

Pondering these unconnected thoughts, Mr Clarke reports that it *is possible*, in principle, to recapture past sounds. But this may only be done when they have been accidentally frozen by some natural or artificial process; otherwise our words are lost for ever by thermal agitation in the air, and their energy sinks below that of random molecules of air.

But in 1969, a scientist named Richard Woodbridge, in a paper entitled "Acoustic Recordings from Antiquity," reported a fascinating experiment. He played loud music to a canvas while it was being painted, and found that short snatches of it, including a spoken word, could be identified on the paint after it had dried.

Leonardo da Vinci is believed to have used a small orchestra to keep Mona Lisa from getting bored while he was painting her, and so perhaps it will one day be possible to discover what music it was playing by an ingenious examination of the canvas in the Louvre.

If this seems incredible, how much more so did our ancestors find those we now find commonplace? Mr Clarke has unearthed an interesting quotation from a Leipzig newspaper in 1839 commenting on the reported invention of photography by Louis Daguerre. This gem should amuse owners of Instamatics:

"The wish to capture evanescent reflections is not only impossible, as has been shown by thorough German investigation, but the will to do so is blasphemy. God created man in his own image, and no man-made machine may fix the image of God. One may straightway call the Frenchman Daguerre, who boasts of such unheard-of things, the fool of fools."

The evanescent shapes which Daguerre detected on his primitive film reminded Mr Clarke of the great vision of Tennyson which inspired his poem "In Memoriam," written in 1850. Very few critics have realised that the subject of the poem is the poet's speculative feelings about the theory of continental drift, which did not become respectable until it was established by Alfred Wegener in 1912.

There rolls the deep where grew the tree.
O earth, what changes hast thou seen!
There where the long street roars hath been.
The stillness of the central sea.

The hills are shadows, and they flow
From form to form, and nothing stands;
They melt like mist, the solid lands,
Like clouds they shape themselves and go.

But how Tennyson was able to describe so explicitly a theory which did not become part of established science until 62 years later I am not altogether clear.

The British Cabinet Minister accused of idiocy was, needless to say, Mr Wedgwood Benn. His sin, which struck at Mr Clarke's entire philosophy of life, was to have held up the first sample of North Sea oil to come ashore and boast that the enterprise was more important than the moon landings "which only brought back soil and rock."

Since North Sea oil will have run out within a couple of decades, while the moon will be for ever remembered as man's gateway to the stars, one feels some sympathy for Mr Clarke's complaint against "some unimaginative politicians whose attention span seldom extends beyond the next Cabinet reshuffle."

To the three great inventions of the Middle Ages which accelerated progress so greatly, namely the compass, gunpowder and the printing press, Mr Clarke adds a fourth—that of spectacles in about 1350, which virtually doubled the intellectual capacity of the human race. A man no longer had to give up reading when he reached his most productive age.

This has been little more than a glance at one of the most delightful books of the year, which gives a deceptive impression of lightness by rambling so discursively through a multitude of scientific subjects.

The author concludes with his own answer to the ancient question: what is the purpose of life? Its purpose, he suggests, is information processing. In other words, to receive information constantly is one of the strongest of human passions.

This would explain, he believes, why a seemingly extravagant idea like the use of communication by synchronous space satellite became an everyday reality 20 years after he himself had first proposed the idea in 1945, while many less ambitious scientific projects have yet to be accomplished.

"The purpose of life is information processing," Mr Clarke says. "Indeed, you may even retort: 'Well, what is the purpose of information processing?' I'm glad you asked me that."

Is an ice age due?

WHAT WILL the earth's climate be like a. half-century from now? A great deal of alarm has been expressed about the prospect, with one scientist using the evidence of pre-historic tree pollens to predict that a full-scale ice age could begin in less than 20 years.

In October, 1979, Dr Genevieve Woillard, of the Catholic University of Louvain, Belgium, reported in *Nature* that she had reached this conclusion after examining the pollens of hardwood trees preserved in different mud layers beneath Grand Pile lake, Vosges.

These hardwood trees, typical of a temperate zone, covered central Europe about 120,000 years ago, the time when the last ice age began.

And how swiftly that ice age began, if we accept Dr Woillard's conclusions. In less than two decades (according to her identification of the different mud layers) the hardwoods had been replaced by forests of pine, spruce and birch now found in Scandinavia, and which are common in the cold landscapes of the north. The implication was clear; a climate of ice and snow had rushed upon Europe with terrifying speed, and age-long winter had descended.

This frigid epoch, during which most of what is now England was over-run with glaciers, and the sites of several modern North American cities were buried a mile under ice, lasted some 100,000 years. Then, some 15,000 years ago, the glaciers retreated, and the present "interglacial" began.

There is an ominous reason why the last 15,000 years, which contains all of recorded human history, should be called an "interglacial." It is because a new ice age is expected to bring it to an end. When will this happen? We don't know precisely, but there is one clue. The duration of the previous interglacial was —wait for it—about 15,000 years. As Dr Woillard puts it, "we cannot exclude the possibility that we already live at the beginning of a terminal interglacial period."

But how much of this is true "evidence" that an ice age is imminent? By itself it proves nothing, but there are further pieces of "evidence" which seem to infer that the northern hermisphere is cooling rapidly. Since about 1940, on several occasions, there have been successive cold winters of unusual severity. So what does this prove? Only, perhaps, that there has been an ordinary run of cold winters.

Mr David Houghton, a specialist on climate at the Meteorological Office at Bracknell, has analysed the records of all runs of cold and mild winters from the beginning of the 19th century to the present in search of some tell-tale pattern. He failed to find one.

But surely, it will be argued, in the 17th and 18th centuries, Europe went through a "mini ice age," in which during many winters people were able to roast oxen on the frozen Thames, as a famous engraving by Wenceslas Hollar testifies. Certainly this happened, but was it because of extreme cold, or merely because London in those days lacked modern industry and shipping which today warms the capital and prevents the Thames from freezing? The construction of the Thames embankments also narrowed the river and speeded its flow, making ice formation more difficult. The science of climatology is like a history of maybes and might-have-beens. The more one investigates it, the less one finds firm evidence of anything.

It is not generally realised that the polar ice-caps are the *least* important region of the world for playing a part in shaping the climate. About 70 per cent of the sun's heat that strikes the lands of ice and snow is reflected back into space. In contrast, the equatorial rain forests and the tropical oceans *absorb* respectively 93 per cent, and 96 per cent, of solar heat.

Anyone, therefore, who predicts a rapidly

approaching ice age on the basis of past events or on any other grounds is using evidence which is only circumstantial. From all the expensive climatic studies which have been undertaken, nothing definite has emerged.

Saving London from the flood

LOW AIR PRESSURE over the North Sea combines with strong northerly gales. A surge tide builds up off the Norfolk coast. It surges up the Thames during the early morning rush-hour in London.

It pours over the river banks in central London, and into the Underground. Within hours, tens of thousands of people have drowned, electrical systems have been destroyed, power stations are flooded, and much of the economy of southern England is out of action.

Assuming total failure of the present sophisticated warning system, this would be the "worst case" scenario of a disaster which could take place at any time between August and April, between now and 1982, when the mighty anti-flood barrier at Woolwich is expected to be completed.

Many people might argue that the great city of London has survived for nearly 2,000 years living cheek-by-jowl with the North Sea, and since it is still here there can be no great cause for alarm.

But this, unhappily, is not the case. Every century the natural defences of London become weaker and every century its proximity to the vast ocean mass of the North Sea becomes more menacing.

For the whole of south-east England is gradually sinking at a rate of about one foot every 100 years. Within a few thousand years, unless some future engineering scheme can prevent it, the new ocean coastline will be a line stretching roughly from South-ampton to the Wash.

This sinking explains why every surge tide running up the Thames since records began has tended to be more destructive than the last.

"Men did row with wherries (boats) in Westminster Hall," reported a chronicler in 1236. "There was last night the greatest tide that was ever remembered in England, all Whitehall having been drowned", Samuel Pepys recorded in his diary in December, 1663.

But these floods were trickles compared with the disaster of Sunday Feb 1, 1953, when Canvey Island, at the northern mouth of the Thames, was over-whelmed by an eight-foot flood. Two-hundred-and-eighty-five people on the east coast were drowned.

In January 1978 a surge came within 12 inches of the point at which emergency sirens would have sounded to warn people to prepare to evacuate London.

The next great flood, when it comes, may be higher still. Ominously, the 1953 flood was a foot higher than the one in 1928, which in turn was 12 inches higher than the great surge of 1881.

The barrier at Woolwich will put an end to this menace for about 60 years, when still more formidable defences will have to be built.

This ingenious contraption, which is costing a little over £200 million, consists of a series of gates, which stay open for shipping in normal weather, but which can shut off the upper Thames completely within 30 minutes in the event of a tidal surge.

The system will work something like this: consider a massive object shaped like a half-cylinder lying in a cradle at the bottom of the river. Its curved part faces downwards and its flat part upwards, so the ships can pass over it.

The flood approaches. The barrier goes into action. The half-cylinder rotates upwards so that it sticks out of the water. The curved part now faces the onrushing mass of water, and stops it.

Nine of these moveable gates, weighing a total of 51,000 tons will comprise the Woolwich barrier, which, is now, apart from nuclear power stations, the biggest civil engineering project under way in Britain.

A peculiarly horrifying and most readable account of what could happen if the Woolwich barrier is *not* built is given in Richard Doyle's recent novel "Deluge" (Pan paperbacks) which, with sensational exaggeration for purposes of fiction, has a death-toll in central London of 150,000. If the next big flood comes before 1982, people will demand to know why it took 29 years to do something about it after the 1953 warning.

Inventions choked by red tape

MY critical remarks about President Carter's absence of a space policy provoked one heated response. It was from an American engineer working in London, who professed to be "appalled" by my "exhalation." At the risk of angering this gentleman further, I must discuss America's dangerous "innovation recession" which has taken a strong turn for the worse under the Carter Administration.

The facts are quite plain. In the nineteen-fifties, America made the largest contributions to the world's technological inventions, just as Britain had done a century earlier. But now, in the late 'seventies, this situation has changed, and by far the greatest number of contributions is being made by West Germany and Japan.

Between 1953 and 1955, the number of US patents granted (a key measure of research and development vitality) was 46,000 for "major" American technological innovations, and 10,000 for non-American ones.

But in 1976, these numbers had changed markedly.

The number of American patents was 33,000, and non-American patents had nearly doubled to 19,000.

In shops and homes, the change has become obvious, Mr Vermont Royster, in a recent article in the *Wall Street Journal,* tells of his surprise and dismay in discovering after a casual inspection of his gadget-filled home, what a large number of his possessions were foreign-made.

"Of the two cars in our garage," he says, "only one (the older one) is American-made. Of our two TV sets, one bears a US trademark but much of its insides are imported; the other bears a foreign label.

"Of three cameras, only one is US made. The two calculators, the two radios, the cassette recorder and the binoculars, all come from overseas. So do the newly-added gadgets to tell us when the house is on fire. Of three typewriters (a journalistic extravagance), two are of foreign manufacture.

"Combing the kitchen appliances, I discovered that neither the electric ice-crusher, nor the cooking-timer nor the popcorn-popper are marked 'Made in the USA.' We seem to have unwittingly stocked an international warehouse."

In Britain, of course, the situation is very much worse. Anyone who goes out shopping for gadgets determined to buy only British will return home with a lean basket. Even in central London, I have been unable to find a single British typewriter or camera.

But to return to America, where the commercial vitality of inventors is still much greater, Mr Arthur M. Bueche, senior vice-president of research and development at General Electric, attributes the decline to a change in the American character.

"We've gone from an expansive gung-ho attitude to a defensive 'What's in it for me?' attitude," he says. "Faced with a challenge, Americans are now likely to say: 'Let's not risk it.'"

But this explanation seems very un-Darwinian. As any good psychologist will know, a nation of 220

million people does not "change its character" in a
mere 20 years.

The true cause must lie nearer at hand; with a
succession of careless governments, but in particular
with the proliferation of meddlesome bureaucrats
under President Carter.

American research scientists sometimes spend more
time dealing with government red tape than in working
in their laboratories. A research contract in energy, for
example, requires no fewer than seven separate
examinations before it can be approved by the Department of Energy.

In a fumbling attempt to combat this trend, Mr
Carter has set up a panel of experts who will probably
recommend the establishment of a permanent council
to provide funds for commercial R and D, rather like
Britain's National Research and Development
Council.

In an ideal world there would be no need for these
absurd councils. Companies invest in new inventions
when they can afford to. And they increasingly *cannot*
afford to when they have to pay excessive taxes to
provide the salaries of ever more numerous bureaucrats. Cut the taxes and sack the bureaucrats and
inventive vitality will revive like the phoenix.

The ultimate reference library

ISAAC ASIMOV once wrote a series of short stories
about a gigantic computer called Multivac, which
stored in its memory banks all the information that
was known to the human race. This dream is now
being gropingly realised by an organisation based in
Philadelphia called the Institute for Scientific Information.

Unlike Multivac, this Institute does not communicate with the public through computer terminals, but rather through a collection of massive books, each about twice as fat as a telephone directory, that are available in most scientific libraries in Britain and elsewhere.

To give a general idea of the scope of the information which they provide, I will say that the number of *pages* they produce each year is about 1 · 1 million, and they are compiled after a million articles from scientific journals have been screened, coded and cross-indexed.

What is the purpose of this seemingly maniacal accumulation of data? Well, back in the 18th century there were only 10 scientific journals, and a single day's reading could bring one completely up to date. By 1830 the number of titles had swelled to 300, and today, so great has been the so-called "information explosion," there are more than 50,000.

It is plainly impossible for a single human being to digest such mountainous quantities of information, and so anyone embarking on a scientific study will be well advised first to consult these volumes.

Let us suppose, to take an imaginary example, that a student of mathematics wants to know whether the circle can be squared. He looks up the word "circle" in the current Subject Index and finds the entry "CIRCLE, squaring . . . Snodgrass, J. M."

He now turns to the Source Index and looks up J. M. Snodgrass. Here he finds the entry: "Snodgrass J. M. A Theoretical proof that the circle can be squared. Rur J. Math 5 92-96 73." This tells him that Prof. Snodgrass's article appeared on pages 92-96 of volume 5 of the *Ruritanian Journal of Mathematics* in 1973. He has now identified the article and knows where to find it.

But the student wonders if Prof. Snodgrass is not something of a crackpot. What do other scientists think of his work? He looks up Snodgrass in the Citation Index, and finds the entry: "Snodgrass J. M.

Rur J Math 5 92-96 73. Tupman P. J. J Clin Psych 23 23-26 74.''

The Source Index quickly identifies the article by P. J. Tupman in the *Journal of Clinical Psychology*, ''Unusual Obsession by a Mathematician: A Case Study of a Delusion Caused by Stress.'' The student's question is now answered: it may be impossible to square the circle, since the only man who claims to have done it is in a lunatic asylum.

The Citation Index is a truly remarkable book. To put it bluntly, it can give a rough idea of whether a particular scientist is any good at his job at all, and as such it is being increasingly used by scientific administrators, those bureaucrats who decide whether a project shall receive a grant or whether a scientist deserves a university post.

It works like this. If a scientific article is cited many times in other articles, it indicates that it has made a great impact and that its author has earned considerable distinction. But if, after a reasonable time period, the article is barely cited at all and then only in a derogatory manner, like that of the wretched Prof. Snodgrass, it tends to show that the scientist in question is not being taken seriously by his colleagues.

Frequency of citation is of course only a rough-and-ready check, since unscrupulous scientists can get round it by the ''buddy system,'' in which they agree privately to cite each other's papers in order to boost their mutual reputations.

The Institute, which already has extra books to cover the social sciences, plans soon to produce books on the arts and humanities. For one thing, this will revolutionise historical research since *all scientific articles which could have any relevance to history will be cross-indexed to the arts and humanities Citation Index.*

Take the case of an author who might be researching for a biography of Henry VIII. While he is busy analysing Tudor politics, a medical scientist diagnoses the king's lingering disease from a detail in one of the Holbein portraits, and reports it in a medical journal. The historian sees the citation and gets a new insight into Henry's personality.

Details of the location of the books in British libraries may be had from 132 High Street, Uxbridge, the Institute's London office.

The definitive theory of ghosts

All argument is against it, but all belief is for it.
　　　—Dr Johnson on the existence of ghosts.

Do ghosts exist? The question can be relied upon to start an argument: "I once saw a ghost"—"After how many drinks?"—"I tell you, I swear I did . . ." And so forth.

But the question can never be settled by this kind of I-did-you-can't-have approach. So let us take a fairly cool scientific look at the ghost problem, and see if by adopting a practical attitude we can come any nearer to resolving it.

If we accept for the sake of argument the hypothesis that ghosts do exist, then we must assume that their behaviour is bound by certain laws. Everything else in the universe is subject to natural laws, and if ghosts share that universe with us, they must be similarly governed.

The first assumption we must make is that every ghost weighs something, that ghosts have mass. This is obvious if one thinks about it. For a ghost to be detectable, it must carry a message. The message can either be simple, where the ghost merely waves a skinny hand and disappears, or it can be complex: "I am thy father's spirit, Hamlet."

And so, in a sense, a ghost is no different from a

radio or TV signal, which also carries a message. Like these electro-magnetic signals, ghosts will consist of charged particles which have mass, and like them also, they will be unable to travel faster than light.

Now since ghosts have mass, we must conclude that their number is finite. With an infinite number of ghosts, each possessing a small amount of mass, the surface of the world would be so intolerably crowded with ghosts, or "ghost-dense," that there would be no room on it for the living.

This plainly is not the case, and the observation enables us to write a few elementary rules which must govern the numbers of ghosts.

It is probable that no dead person can produce more than one ghost, and that no ghost can be in more than one place at one time. If either of these were not the case, a malevolent dead person could unleash dozens of identical ghost-clones with which to haunt his enemies simultaneously while they were in different places. No psychic researcher has ever reported such a phenomenon. (It is significant that the several ghosts who appeared to Richard III and Richmond before the Battle of Bosworth in Shakespeare's play spoke to each man in turn and not to both at once.) And so ghosts cannot duplicate themselves.

Are ghosts immortal? This seems most unlikely. Spiritual immortality would violate the Second Law of Thermodynamics, which predicts that every organised system must eventually break down. Ghosts may have long life-spans compared with ourselves—the ghost of Charles I who died 300 years ago is still sometimes reported—but like a radio signal which fades with distance, they must become undetectable beyond a period which no doubt partly depends on the strength of their personalities and their desire to haunt people.

Armed with these proofs, we can write the definitive Ghosts Equation, which determines the number of ghosts in the world:

$N(g)$ equals or is less than P.

Where P is the present human population, and $N(g)$ is the present number of ghosts. It is axiomatic that no living person can produce a ghost (at least not until electronics is considerably more developed), and that the world population is now so great that it equals the total number of people who have ever lived and died.

Now the Second Law of Thermodynamics only predicts probabilities and not certainties, and one might think it possible that a few ghosts of ape-men and of earlier animal ancestors of man might have swollen the 4,200 million ghost population.

But this does not seem very likely. One would expect the spirits to share some of our habits, including that of polite social intercourse. After all, what does a self-respecting ghost do when it is not clanking its chain in a cellar? One assumes that there are periodic ghostly get-togethers.

But the ape-men were illiterate. They would be incapable of joining an intelligent discussion about politics or literature, and they would probably smell. It seems likely therefore that the Ghost Equation is correct and complete, since such inferior phantoms as the wraiths of primeval apes would never be admitted into the drawing rooms of ghostly society. Physical state may change, but snobbery is eternal.

Could we not put our presumed knowledge of ghosts to some practical use? Since they can travel at the speed of light, could we not send them as messengers to alien civilisations on distant planets? After all, a typical ghost is already conveniently encoded with several million bytes of information.

But it might after all be cheaper and less bother to use radio. Ghosts can be mischievous and bad-mannered. The aliens might take offence if terrestial ambassadors behaved like the terrifying entities in the short stories of M. R. James, or if they rudely occupied the seats of their hosts at dinner-parties, like the shade of Banquo.

INDEX